ORGANISATION DE COOPÉRATION ET DE DÉVELOPPEMENT ÉCONOMIQUES

En vertu de l'article 1er de la Convention signée le 14 décembre 1960, à Paris, et entrée en vigueur le 30 septembre 1961, l'Organisation de coopération et de développement économiques (OCDE) a pour objectif de promouvoir des politiques visant :

– à réaliser la plus forte expansion de l'économie et de l'emploi et une progression du niveau de vie dans les pays Membres, tout en maintenant la stabilité financière, et à contribuer ainsi au développement de l'économie mondiale ;

– à contribuer à une saine expansion économique dans les pays Membres, ainsi que les pays non membres, en voie de développement économique ;

– à contribuer à l'expansion du commerce mondial sur une base multilatérale et non discriminatoire conformément aux obligations internationales.

Les pays Membres originaires de l'OCDE sont : l'Allemagne, l'Autriche, la Belgique, le Canada, le Danemark, l'Espagne, les États-Unis, la France, la Grèce, l'Irlande, l'Islande, l'Italie, le Luxembourg, la Norvège, les Pays-Bas, le Portugal, le Royaume-Uni, la Suède, la Suisse et la Turquie. Les pays suivants sont ultérieurement devenus Membres par adhésion aux dates indiquées ci-après : le Japon (28 avril 1964), la Finlande (28 janvier 1969), l'Australie (7 juin 1971), la Nouvelle-Zélande (29 mai 1973), le Mexique (18 mai 1994), la République tchèque (21 décembre 1995), la Hongrie (7 mai 1996), la Pologne (22 novembre 1996) et la Corée (12 décembre 1996). La Commission des Communautés européennes participe aux travaux de l'OCDE (article 13 de la Convention de l'OCDE).

L'AGENCE DE L'OCDE POUR L'ÉNERGIE NUCLÉAIRE

L'Agence de l'OCDE pour l'énergie nucléaire (AEN) a été créée le 1er février 1958 sous le nom d'Agence européenne pour l'énergie nucléaire de l'OECE. Elle a pris sa dénomination actuelle le 20 avril 1972, lorsque le Japon est devenu son premier pays Membre de plein exercice non européen. L'Agence groupe aujourd'hui tous les pays Membres de l'OCDE, à l'exception de la Nouvelle-Zélande et de la Pologne. La Commission des Communautés européennes participe à ses travaux.

L'AEN a pour principal objectif de promouvoir la coopération entre les gouvernements de ses pays participants pour le développement de l'énergie nucléaire en tant que source d'énergie sûre, acceptable du point de vue de l'environnement et économique.

Pour atteindre cet objectif, l'AEN :

– *encourage l'harmonisation des politiques et pratiques réglementaires notamment en ce qui concerne la sûreté des installations nucléaires, la protection de l'homme contre les rayonnements ionisants et la préservation de l'environnement, la gestion des déchets radioactifs, ainsi que la responsabilité civile et l'assurance en matière nucléaire ;*

– *évalue la contribution de l'électronucléaire aux approvisionnements en énergie, en examinant régulièrement les aspects économiques et techniques de la croissance de l'énergie nucléaire et en établissant des prévisions concernant l'offre et la demande de services pour les différentes phases du cycle du combustible nucléaire;*

– *développe les échanges d'information scientifiques et techniques notamment par l'intermédiaire de services communs ;*

– *met sur pied des programmes internationaux de recherche et développement, et des entreprises communes.*

Pour ces activités, ainsi que pour d'autres travaux connexes, l'AEN collabore étroitement avec l'Agence internationale de l'énergie atomique de Vienne, avec laquelle elle a conclu un Accord de coopération, ainsi qu'avec d'autres organisations internationales opérant dans le domaine nucléaire.

NUCLEAR ENERGY AGENCY
ORGANISATION FOR ECONOMIC CO-OPERATION AND DEVELOPMENT

Glossary
of Nuclear Power Plant Ageing

Glossaire du vieillissement
des centrales nucléaires

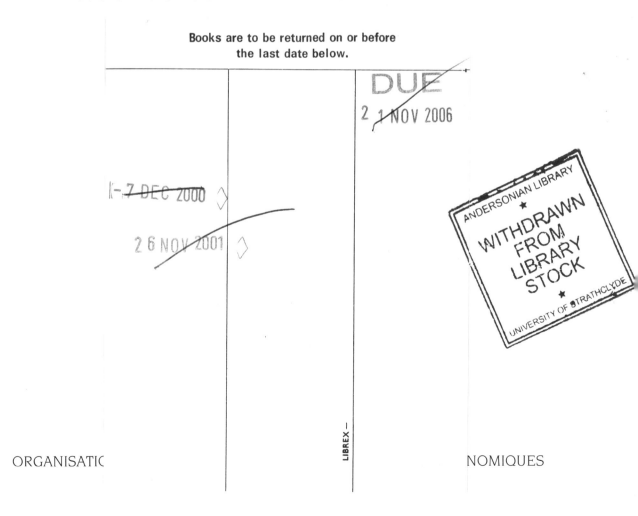
ORGANISATIC   NOMIQUES

ORGANISATION FOR ECONOMIC CO-OPERATION AND DEVELOPMENT

Pursuant to Article 1 of the Convention signed in Paris on 14th December 1960, and which came into force on 30th September 1961, the Organisation for Economic Co-operation and Development (OECD) shall promote policies designed:

- to achieve the highest sustainable economic growth and employment and a rising standard of living in Member countries, while maintaining financial stability, and thus to contribute to the development of the world economy;
- to contribute to sound economic expansion in Member as well as non-member countries in the process of economic development; and
- to contribute to the expansion of world trade on a multilateral, non-discriminatory basis in accordance with international obligations.

The original Member countries of the OECD are Austria, Belgium, Canada, Denmark, France, Germany, Greece, Iceland, Ireland, Italy, Luxembourg, the Netherlands, Norway, Portugal, Spain, Sweden, Switzerland, Turkey, the United Kingdom and the United States. The following countries became Members subsequently through accession at the dates indicated hereafter; Japan (28th April 1964), Finland (28th January 1969), Australia (7th June 1971), New Zealand (29th May 1973), Mexico (18th May 1994), the Czech Republic (21st December 1995), Hungary (7th May 1996), Poland (22nd November 1996) and the Republic of Korea (12th December 1996). The Commission of the European Communities takes part in the work of the OECD (Article 13 of the OECD Convention).

NUCLEAR ENERGY AGENCY

The OECD Nuclear Energy Agency (NEA) was established on 1st February 1958 under the name of OEEC European Nuclear Energy Agency. It received its present designation on 20th April 1972, when Japan became its first non-European full Member. NEA membership today consist of all OECD Member countries, except New Zealand and Poland. The Commission of the European Communities takes part in the work of the Agency.

The primary objective of the NEA is to promote co-operation among the governments of its participating countries in furthering the development of nuclear power as a safe, environmentally acceptable and economic energy source.

This is achieved by:

- *encouraging harmonization of national regulatory policies and practices, with particular reference to the safety of nuclear installations, protection of man against ionising radiation and preservation of the environment, radioactive waste management, and nuclear third party liability and insurance;*
- *assessing the contribution of nuclear power to the overall energy supply by keeping under review the technical and economic aspects of nuclear power growth and forecasting demand and supply for the different phases of the nuclear fuel cycle;*
- *developing exchanges of scientific and technical information particularly through participation in common services;*
- *setting up international research and development programmes and joint undertakings.*

In these and related tasks, the NEA works in close collaboration with the International Atomic Energy Agency in Vienna, with which it has concluded a Co-operation Agreement, as well as with other international organisations in the nuclear field.

GLOSSARY OF NUCLEAR POWER PLANT AGEING

A glossary useful for understanding and managing

the ageing of nuclear power plant systems,

structures and components

WHY COMMON AGEING TERMINOLOGY?

As the service life of operating nuclear power plants increases, potential misunderstanding of terms related to ageing degradation of Systems, Structures, or Components (SSCs) is receiving more attention. Common ageing terminology has been developed to improve the understanding of ageing phenomena, facilitate the reporting of relevant plant failure data, and promote uniform interpretations of standards and regulations that address ageing.

The terminology should be useful in the areas of life and ageing management. Life management can minimise operations and maintenance costs and can support the option of extending the operating term of a plant from 40 to 60 years. More importantly, effective ageing management contributes to the maintenance of adequate plant safety margins.

Recognising the importance of clear communication in these areas, representatives from a cross-section of the industry have developed a uniform vocabulary of terms relating to ageing.

In view of the benefits to be gained, use of common ageing terminology is recommended in technical documents, failure reports, research reports, future regulations, and other documentation related to nuclear power plant ageing. Documentation would state that common ageing terminology is used except where noted. Appropriate exceptions would be cases for which the writer opts to use definitions from existing standards and regulations.

The main benefits from the use of common ageing terminology are:

- improved reporting and interpretation of plant data on SSC degradation and failure, including accurate root cause identification;

- improved interpretation and compliance with codes, standards, and regulations related to nuclear plant ageing.

Why this glossary?

The Nuclear Energy Agency (NEA) has published this glossary, in co-operation with the Commission of the European Communities (CEC) and the International Atomic Energy Agency (IAEA), as a handy reference to facilitate and encourage widespread use of common ageing terminology. The goal is to provide plant personnel (and others who address ageing) with a common set of terms that have uniform, industry-wide meanings, and to facilitate discussion between experts from different countries.

In each language section terms are listed alphabetically, with sequential number. These numbers are repeated in the English language section thus allowing cross-reference between all languages. The dash ("–") in the columns of the reference number indicates that there is no corresponding term in certain language. Note that dashes appear only in the columns of synonymous terms.

In each language the glossary begins with an overview of all terms grouped into six categories. This is followed by an alphabetical listing of terms, definitions, and a few examples. The last pages of each section contain diagrams and a list of key ideas to help illuminate the terminology.

Acknowledgements

This glossary follows closely a publication by the Electric Power Research Institute (EPRI) (BR-101747) to whom we are grateful for support and advice in making available this international version. The EPRI glossary was developed with contributions from several US electric utilities, the Nuclear Energy Institute, the US Nuclear Regulatory Commission, and US national laboratories.

Principal common ageing terms listed by category (synonyms not included)

DEGRADATION	
CAUSES OF DEGRADATION	DEGRADATION/AGEING

Condition
  • Service conditions
  • Pre-service conditions
  • Environmental conditions
  • Functional conditions
  • Operating conditions
  • Normal conditions
  • Error-induced conditions
  • Design basis event
  • Design basis event conditions
  • Design conditions
Stressor
  • Normal stressor
  • Error-induced stressor
  • Design basis event stressor
	Characteristic
	Condition
	  • Degraded condition
	Ageing
	  • Natural ageing
	  • Premature ageing
	  • Normal ageing
	  • Artificial ageing
	  • Accelerated ageing
	  • Age conditioning
	Ageing mechanism
	Ageing effects
	  • Combined effects
	  • Simultaneous effects
	  • Synergistic effects
	Degradation
	  • Ageing degradation
	  • Normal ageing degradation
	  • Error-induced ageing degradation
	Ageing assessment

LIFE CYCLE	
LIFE	FAILURE

Age
Time in service
Life
  • Installed life
  • Service life
  • Remaining life
  • Design life
  • Remaining design life
  • Qualified life
Retirement
	Failure
	  • Degraded failure
	  • Complete failure
	  • Random failure
	  • Common cause failure
	  • Common mode failure
	  • Wearout
	Failure cause
	  • Root cause
	  • Failure mechanism
	  • Failure mode
	Failure analysis
	  • Failure evaluation
	  • Failure modes and effects analysis
	  • Failure trending
	Mean time between failures

AGEING MANAGEMENT	
MAINTENANCE	CONDITION ASSESSMENT

Ageing management
Life management
Maintenance
  • Preventive maintenance
  • Periodic maintenance
  • Planned maintenance
  • Corrective maintenance
Repair
Refurbishment
Overhaul
Replacement
Servicing
Post-maintenance testing
Rework
	Predictive maintenance
	In-service inspection
	In-service test
	Surveillance
	Surveillance requirements
	Condition monitoring
	Condition indicator
	Functional indicator
	Testing
	Diagnosis
	Acceptance criterion

		French	German	Spanish	Russian
1	**Accelerated ageing:** artificial ageing in which the simulation of natural ageing approximates, in a short time, the ageing effects of longer-term service conditions (see also *premature ageing*)	101	34	102	74
2	**Acceptance criterion:** specified limit of a functional or condition indicator used to assess the ability of an SSC[1] to perform its design function	22	3	29	26
3	**Age:** (noun) time from fabrication of an SSC to a stated time	1	4	38	5
4	**Age conditioning:** simulation of natural ageing effects in an SSC by the application of any combination of artificial and natural ageing	60	89	52	29
5	**Ageing:** (noun) general process in which characteristics of an SSC gradually change with time or use	106	5	46	69
6	**Ageing assessment:** evaluation of appropriate information for determining the effects of ageing on the current and future ability of SSCs to function within acceptance criteria	65	9	89	44
7	**Ageing degradation:** ageing effects that could impair the ability of an SSC to function within acceptance criteria • *Examples*: reduction in diameter from wear of a rotating shaft, loss in material strength from fatigue or thermal ageing, swell of potting compounds, and loss of dielectric strength or cracking of insulation	31	7, 8	31	81
8	**Ageing effects:** net changes in characteristics of an SSC that occur with time or use and are due to ageing mechanisms • *Examples*: negative effects – see *ageing degradation*; positive effects – increase in concrete strength from curing; reduced vibration from wear-in of rotating machinery	49	6	40	96
9	**Ageing management:** engineering, operations, and maintenance actions to control within acceptable limits ageing degradation and wearout of SSCs • *Examples* of engineering actions: design, qualification, and failure analysis • *Examples* of operations actions: surveillance, carrying out operational procedures within specified limits, and performing environmental measurements	74	10	61	73

---

1. SSC = system, structure, or component.

		French	German	Spanish	Russian
10	**Ageing mechanism:** specific process that gradually changes characteristics of an SSC with time or use • *Examples*: curing, wear, fatigue, creep, erosion, micro-biological fouling, corrosion, embrittlement, and chemical or biological reactions	86	11	73	28
11	**Age-related degradation:** synonym for *ageing degradation*	30	7, 8	34	–
12	**Artificial ageing:** simulation of natural ageing effects on SSCs by application of stressors representing plant pre-service and service conditions, but perhaps different in intensity, duration, and manner of application		51	48	17
13	**Breakdown:** synonym for *complete failure*	–	79	8	7
14	**Characteristic:** property or attribute of an SSC (such as shape, dimension, weight, condition indicator, functional indicator, performance, or mechanical, chemical, or electrical property)	5	38	9	91
15	**Combined effects:** net changes in characteristics of an SSC produced by two or more stressors	48	50	39	65
16	**Common cause failure:** two or more failures due to a single cause	27	16, 77	57	41
17	**Common mode failure:** two or more failures in the same manner or mode due to a single cause	26	19, 77	58	14
18	**Complete failure:** failure in which there is a complete loss of function	25	79	56	51
19	**Condition (1):** surrounding physical state or influence that can affect an SSC	8	31	12	75
20	**Condition (2):** the state or level of characteristics of an SSC that can affect its ability to perform a design function	–	99	13	66
21	**Condition indicator:** characteristic that can be observed, measured, or trended to infer or directly indicate the current and future ability of an SSC to function within acceptance criteria	75	100	62	50

		French	German	Spanish	Russian
22	**Condition monitoring:** observation, measurement, or trending[2] of condition or functional indicators with respect to some independent parameter (usually time or cycles) to indicate the current and future ability of an SSC to function within acceptance criteria	97	101	28	24
23	**Condition trending:** synonym for *condition monitoring*	57	101	–	56
24	**Corrective maintenance:** actions that restore, by repair, overhaul or replacement, the capability of a failed SSC to function within acceptance criteria	80	49	68	18
25	**Degradation:** immediate or gradual deterioration of characteristics of an SSC that could impair its ability to function within acceptance criteria	29	2, 88	30	80
26	**Degraded condition:** marginally acceptable condition of an unfailed SSC that could lead to a decision to perform planned maintenance	61	1	14	84
27	**Degraded failure:** failure in which a functional indicator does not meet an acceptance criterion, but design function is not completely lost	71	18	59	40
28	**Design basis conditions:** synonym for *design conditions*	–	25	17	–
29	**Design basis event:** event specified to establish reliable performance of normal operating functions and, through deterministic safety analysis, safety-related functions of SSCs; events include anticipated transients, design basis accidents, external events, and natural phenomena	66	26	84	21
30	**Design basis event conditions:** service conditions produced by design basis events	11	32	23	77
31	**Design basis event stressor:** stressor that stems from design basis events and can produce immediate or ageing degradation beyond that produced by normal stressors	68	24	4	87
32	**Design conditions:** service conditions specified for an SSC according to existing rules and guidelines that are normally included in technical specifications (generally includes margin of conservatism beyond expected service conditions)	18	25	18	57

---

2. Discerning a trend in observation or measurements.

		French	German	Spanish	Russian
33	**Design life:** period during which an SSC is expected to function within acceptance criteria	41	27	91	58
34	**Design service conditions:** synonym for *design conditions*	13	25	22	–
35	**Deterioration:** synonym for *degradation*	34	2, 88	36	–
36	**Diagnosis:** examination and evaluation of data to determine either the condition of an SSC or the causes of the condition	35	37	37	9
37	**Diagnostic evaluation:** synonym for *diagnosis*	62	37	–	10
38	**Environmental conditions:** ambient physical states surrounding an SSC • *Examples*: temperature, radiation, and humidity in containment during normal operation or accidents	9	85	15	78
39	**Error-induced ageing degradation:** ageing degradation produced by error-induced conditions	32	13	32	83
40	**Error-induced conditions:** adverse pre-service or service conditions produced by design, fabrication, installation, operation or maintenance errors	17	14, 46	25	76
41	**Error-induced stressor:** stressor that stems from error-induced conditions and can produce immediate or ageing degradation beyond that produced by normal stressors	69	12	2	86
42	**Failure:** inability or interruption of ability of an SSC to function within acceptance criteria	23	15, 86	54	39
43	**Failure analysis:** systematic process of determining and documenting the mode, mechanism, causes, and root cause of failure of an SSC	3	20	6	3
44	**Failure cause:** circumstances during design, manufacture, test or use that have led to failure	6	23	10	54
45	**Failure evaluation:** conclusion drawn from failure analysis	63	21	53	43
46	**Failure mechanism:** physical process that results in failure • *Examples*: cracking of an embrittled cable insulation (ageing-related); an object obstructing flow (non-ageing-related)	85	22	74	27

		French	German	Spanish	Russian
47	**Failure mode:** the manner or state in which an SSC fails	87	87	75	4

- *Examples*: stuck open (valve), short to ground (cable), bearing seizure (motor), leakage (valve, vessel, or containment), flow stoppage (pipe or valve), failure to produce a signal that drops control rods (reactor protection system), and crack or break (structure)

		French	German	Spanish	Russian
48	**Failure modes and effects analysis:** systematic process for determining and documenting potential failure modes and their effects on SSCs	4	42	7	1
49	**Failure trending:** recording, analysing, and extrapolating in-service failures of an SSC with respect to some independent parameter (usually time or cycles)	58	80	86	55
50	**Functional conditions:** influences on an SSC resulting from the performance of design functions (operation of a system or component and loading of a structure)	16	43	24	90

- *Examples*:
  - For a check valve: operational cycling and chatter
  - For a reactor vessel relief valve: reactor coolant pressure, high flow velocities, and temperature increase from the reactor coolant

		French	German	Spanish	Russian
51	**Functional indicator:** condition indicator that is a direct indication of the current ability of an SSC to function within acceptance criteria	77	44	64	89
52	**In-service inspection:** methods and actions for assuring the structural and pressure-retaining integrity of safety-related nuclear power plant components	78	73, 97	66	92
53	**In-service life:** synonym for *service life*	–	41, 65 75	–	–
54	**In-service test:** a test to determine the operational readiness of a component or system	53	96	45	93
55	**Installed life:** period from installation to retirement of an SSC	37	47	94	68
56	**Life:** period from fabrication to retirement of an SSC	36	53	90	13
57	**Life assessment:** synonym for *ageing assessment*	64	54	–	42
58	**Life cycle management:** synonym for *life management*	73	56	–	–

		French	German	Spanish	Russian
59	**Life management:** integration of ageing management and economic planning to: (1) optimise the operation, maintenance, and service life of SSCs; (2) maintain an acceptable level of performance and safety; and (3) maximise return on investment over the service life of the plant	72	55	60	72
60	**Lifetime:** synonym for *life*	100	53	–	–
61	**Maintenance:** aggregate of direct and supporting actions that detect, preclude, or mitigate degradation of a functioning SSC, or restore to an acceptable level the design functions of a failed SSC	79	48, 95	67	71
62	**Malfunction:** synonym for *failure*	28	15, 86	–	32
63	**Mean time between failures:** arithmetic average of operating times between failures of an item	88	57	88	67
64	**Natural ageing:** ageing of an SSC that occurs under pre-service and service conditions, including error-induced conditions	103	60	49	12
65	**Normal ageing:** natural ageing from error-free pre-service or service conditions	104	62	50	33
66	**Normal ageing degradation:** ageing degradation produced by normal conditions	33	63	33	82
67	**Normal conditions:** operating conditions of a properly designed, fabricated, installed, operated, and maintained SSC, excluding design basis event conditions	19	61	26	34
68	**Normal operating conditions:** synonym for *normal conditions*	20	61	20	35
69	**Normal stressor:** stressor that stems from normal conditions and can produce ageing mechanisms and effects in an SSC	70	64	3	88
70	**Operating conditions:** service conditions, including normal and error-induced conditions, prior to the start of a design basis accident or earthquake	12	35	19	95
71	**Operating service conditions:** synonym for *operating conditions*	15	35	–	–

		French	German	Spanish	Russian
72	**Operational conditions:** synonym for *functional conditions*	21	43	27	–
73	**Overhaul:** (noun) extensive repair, refurbishment, or both	96	81	83	22
74	**Performance indicator:** synonym for *functional indicator*	76	44	63	49
75	**Periodic maintenance:** form of preventive maintenance consisting of servicing, parts replacement, surveillance or testing at predetermined intervals of calendar time, operating time or number of cycles	81	66	69	45
76	**Planned maintenance:** form of preventive maintenance consisting of refurbishment or replacement that is scheduled and performed prior to expected unacceptable degradation of an SSC	84	67	70	48
77	**Post-maintenance testing:** testing after maintenance to verify that maintenance was performed correctly and that the SSC can function within acceptance criteria	55	69	44	20
78	**Preconditioning:** synonym for *age conditioning*	59	89	–	–
79	**Predictive maintenance:** form of preventive maintenance performed continuously or at intervals governed by observed condition to monitor, diagnose or trend an SSC's functional or condition indicators; results indicate current and future functional ability or the nature and schedule for planned maintenance	83	90	71	59
80	**Premature ageing:** ageing in service that proceeds at a more rapid rate than expected	105	93	51	53
81	**Pre-service conditions:** actual physical states or influences on an SSC prior to initial operation (e.g. fabrication, storage, transportation, installation, and pre-operational testing)	10	33, 92	16	52
82	**Preventive maintenance:** actions that detect, preclude or mitigate degradation of a functional SSC to sustain or extend its useful life by controlling degradation and failures to an acceptable level; there are three types of preventive maintenance: periodic, predictive and planned	82	91	72	47

		French	German	Spanish	Russian
83	**Qualified life:** period for which an SSC has been demonstrated, through testing, analysis or experience, to be capable of functioning within acceptance criteria during specified operating conditions while retaining the ability to perform its safety functions in a design basis accident or earthquake	40	45	76	79
84	**Random failure:** any failure whose cause or mechanism, or both, make its time of occurrence unpredictable	24	98	55	64
85	**Reconditioning:** synonym for *overhaul*	92	81	–	–
86	**Refurbishment:** planned actions to improve the condition of an unfailed SSC	94	59	78	6
87	**Remaining design life:** period from a stated time to planned retirement of an SSC	42	71	96	37
88	**Remaining life:** actual period from a stated time to retirement of an SSC	44	72	95	38
89	**Remaining service life:** synonym for *remaining life*	39	72	93	–
90	**Remaining useful life:** synonym for *remaining life*	46	72	98	–
91	**Repair:** actions to return a failed SSC to an acceptable condition	95	70	79	61
92	**Replacement:** removal of an undegraded, degraded, or failed SSC or a part thereof and installation of another in its place that can function within the original acceptance criteria	93	29	85	15
93	**Residual life:** synonym for *remaining life*	43	72	–	–
94	**Retirement:** final withdrawal from service of an SSC	90	28	81	8
95	**Rework:** correction of inadequately performed fabrication, installation, or maintenance	91	58	77	11
96	**Root cause:** fundamental reason(s) for an observed condition of an SSC that if corrected prevents recurrence of the condition	7	39	11	25
97	**Root cause analysis:** synonym for *failure analysis*	2	20	5	2

		French	German	Spanish	Russian
98	**Service conditions:** actual physical states or influences during the service life of an SSC, including operating conditions (normal and error-induced), design basis event conditions, and post design basis event conditions	14	**40, 52**	21	60
99	**Service life:** actual period from initial operation to retirement of an SSC	38	41, 65, 75	92	62
100	**Servicing:** routine actions (including cleaning, adjustment, calibration, and replacement of consumables) that sustain or extend the useful life of an SSC	52	94	82	70
101	**Simultaneous effect:** combined effects from stressors acting simulataneously	50	74	42	36
102	**Stressor:** agent or stimulus that stems from pre-service and service conditions and can produce immediate or ageing degradation of an SSC • *Examples*: heat, radiation, humidity, steam, chemicals, pressure, vibration, seismic motion, electrical cycling, and mechanical cycling	67	30	1	85
103	**Surveillance (1):** observation or measurement of condition or functional indicators to verify that an SSC currently can function within acceptance criteria	98	82	65	94
104	**Surveillance (2)[3]:** continuous monitoring of plant conditions during operation or shut down	–	83	99	23
105	**Surveillance requirements:** test, calibration, or inspection to assure that the necessary quality of systems and components is maintained, that facility operation will be within the safety limits, and that the limiting conditions of operation will be met (use only when specific regulatory and legal connotations are called for)	89	84	80	31
106	**Surveillance testing:** synonym for *surveillance, surveillance requirements*, and *testing* (use only when specific regulatory and legal connotations are called for)	56	–	–	30
107	**Synergistic effects:** portion of changes in characteristics of an SSC produced solely by the interaction of stressors acting simultaneously, as distinguished from changes produced by superposition from each stressor acting independently	51	76	41	63

---

3. Some people use this term for the meaning indicated here, which in some texts may be called "in-service inspection", though neither is common usage in the USA.

		French	German	Spanish	Russian
108	**Testing:** observation or measurement of condition indicators under controlled conditions to verify that an SSC currently conforms to acceptance criteria	54	68, 78	43	19
109	**Time in service:** time from initial operation of an SSC to a stated time	47	36	87	46
110	**Useful life:** synonym for *service life*	45	41, 65, 75	97	–
111	**Wearout:** failure produced by an ageing mechanism	99	17	35	16

KEY IDEAS OF COMMON AGEING TERMINOLOGY

Causes of degradation

- **Service conditions** are all actual conditions that influence an SSC. They encompass **operating conditions** (including **normal** and **error-induced conditions** as well as anticipated transients) and accident conditions.

- **Design conditions** are hypothetical conditions generally specified to include a margin of conservatism beyond expected actual service conditions.

Life

- **Service life** is the actual period an SSC provides useful service. This may differ from the expected service life, i.e. **design life**.

- The **age** of an SSC (measured from its time of fabrication) may differ from its **time in service** (measured from initial operation of the SSC).

Degradation/Ageing

- **Degradation** is gradual (ageing) or immediate (non-ageing).

- **Ageing degradation** is produced by **operating conditions**, including both **environmental conditions** such as temperature and radiation as well as **functional conditions** such as relative motion between parts. Operating conditions produce **normal stressors** or **error-induced stressors**.

- **Design basis events** include anticipated transients during plant operation (which can contribute to **ageing**) and design basis accidents and earthquakes (which produce immediate, not ageing, **degradation**).

- A **degraded condition** is marginally acceptable, but is not a **failure**. In a **degraded failure**, **acceptance criteria** on condition or performance are not met, but there is partial function. In **complete failure** there is no function.

Failure

- **Failure** is usually produced by a sequential chain of causes, not a single cause. **Wearout** is a failure whose last cause is an **ageing mechanism**. The **root cause** may not be that ageing mechanism.

- **Premature ageing** may cause in-service **failure** of an SSC. The term **accelerated ageing** should be reserved for **artificial ageing**, usually performed in a laboratory.

- **Failure analyses** identify **failure causes**, the **failure mechanism**, and the **failure mode**. Each of these terms has a different meaning. The **root cause** of **error-induced ageing degradation** and failures is not **ageing**, but rather human error.

Maintenance/Condition assessment

- **Maintenance** is a broad term that includes **corrective maintenance** and **preventive maintenance**. **Maintenance** may be performed by maintenance, engineering, or operations personnel. **Preventive maintenance** includes **predictive maintenance** such as **surveillance, testing**, and **condition monitoring**.

- **Repair** is performed only on a failed SSC; **refurbishment** is performed only on an unfailed SSC. An **overhaul** is an extensive repair and/or refurbishment.

Figure 1. **Relationships among term categories**

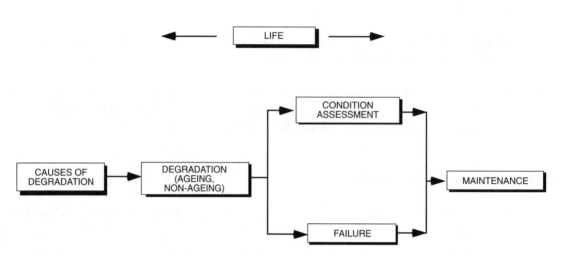

Note: Condition assessment is actually the predictive part of maintenance.

Figure 2. Relationships among types of service conditions

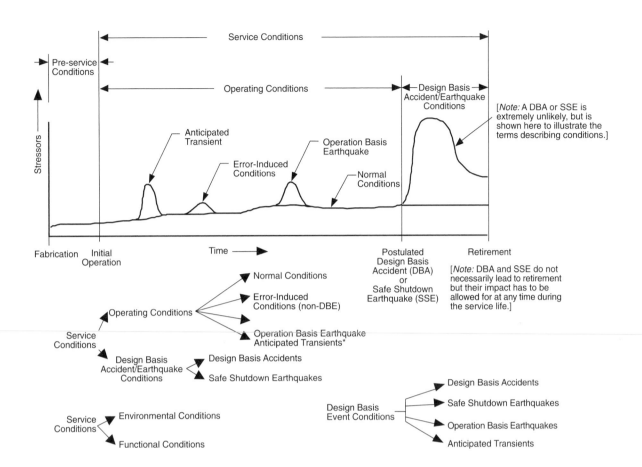

* Note that anticipated transients, although designed for, may be induced by error. Design conditions include all service conditions except non-DBE error-induced conditions.

Figure 3. **Relationships among ageing terms**

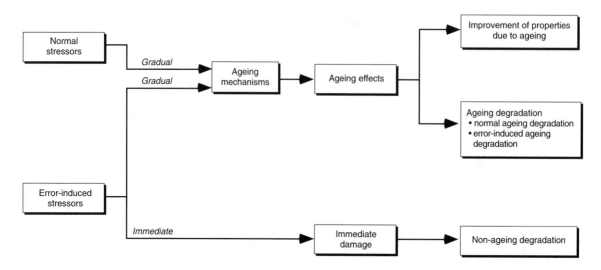

Figure 4. **Relationships among terms describing actual life events on an event timeline**

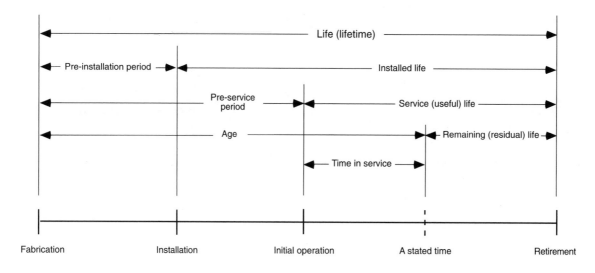

Note: These terms define actual milestones in the past or future history of a system, structure, or component. Synonymous terms are shown in parentheses.

Figure 5.   **Relationships among terms describing predictions of design life**

A.   Initial design life prediction

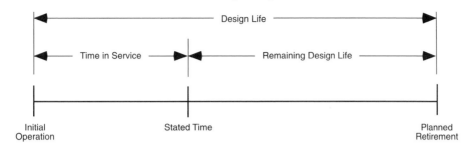

B.   Possible longer design life due to revised design life prediction (discretionary)

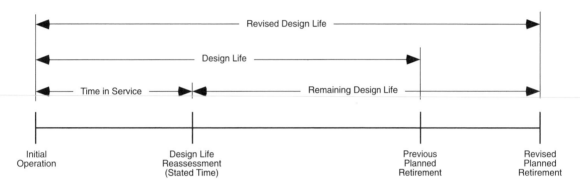

C.   Possible shorter design life due to revised design life prediction (discretionary)

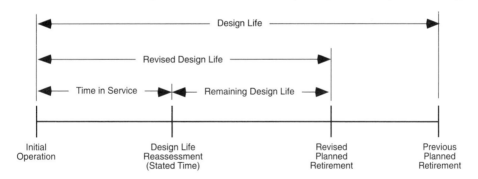

Figure 6.   **Relationships among failure terms**

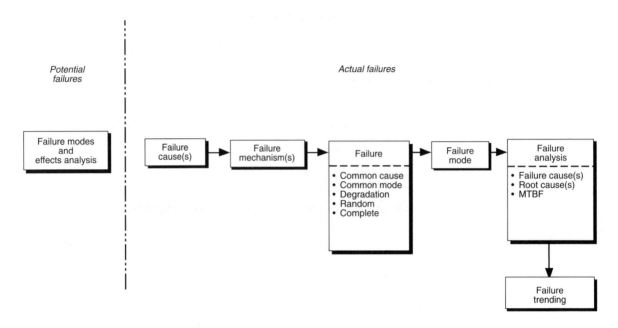

Figure 7.   **Examples of causes, mechanism, and modes**

Definitions of root causes and wearout

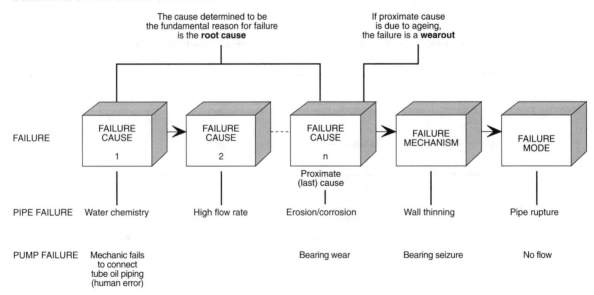

Figure 8.   **Terms describing maintenance**

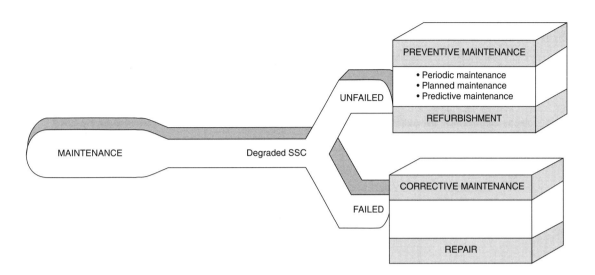

Figure 9. **Relationships among maintenance terms**

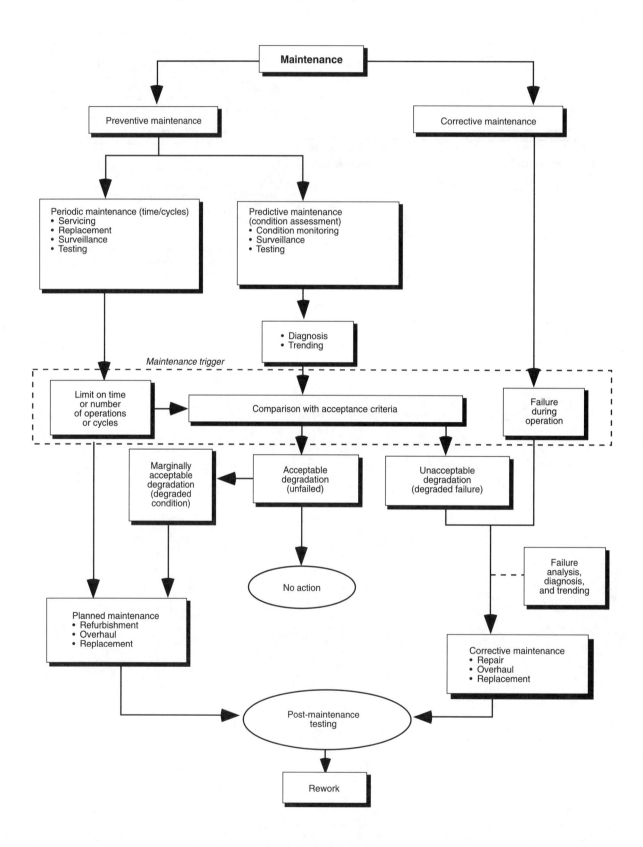

GLOSSAIRE DU VIEILLISSEMENT DES CENTRALES NUCLÉAIRES

Un glossaire utile pour comprendre et gérer le vieillissement
des systèmes, structures et composants des centrales nucléaires

POURQUOI UNE TERMINOLOGIE COMMUNE DU VIEILLISSEMENT ?

Comme la durée de vie utile des centrales nucléaires en exploitation augmente, on prête davantage d'attention aux malentendus possibles sur la terminologie de la dégradation par vieillissement des systèmes, des structures ou des composants (SSC). Cette terminologie commune du vieillissement a été préparée pour améliorer la compréhension des phénomènes de vieillissement, faciliter la communication de données pertinentes sur les défaillances des centrales et permettre une interprétation uniforme des normes et des règlements sur le vieillissement.

Elle devrait être utile dans le domaine de la gestion de la durée de vie et du vieillissement. La gestion de la durée de vie peut permettre de réduire au minimum les coûts d'exploitation et de maintenance et peut apporter des arguments en faveur de la prolongation de l'exploitation d'une centrale de 40 à 60 ans. En outre, et ce qui est bien plus important, une gestion du vieillissement efficace contribue à maintenir des marges de sécurité adéquates dans la centrale.

Conscients de l'importance d'une communication dénuée de toute ambiguïté dans ces domaines, les représentants d'un échantillon d'activités de l'industrie nucléaire ont établi un glossaire uniforme de termes se rapportant au vieillissement.

Il est recommandé d'utiliser cette terminologie du vieillissement, en raison des bénéfices que l'on pourra en tirer, dans les documents techniques, les rapports sur les défaillances, les rapports de recherche, les futurs règlements ainsi que tout autre document se rapportant au vieillissement des centrales nucléaires. Il conviendra(it) alors d'indiquer dans les documents que la terminologie commune du vieillissement y est utilisée excepté aux endroits mentionnés. Cette terminologie pourra ne pas être utilisée lorsque l'auteur choisit d'employer les définitions apparaissant dans des normes ou des règlements déjà existants.

L'utilisation d'une terminologie commune du vieillissement présente les avantages suivants :

- amélioration de la communication et de l'interprétation des données sur la dégradation et la défaillance des SSC des centrales, y compris le diagnostic exact de la cause profonde ;

- amélioration de l'interprétation et de l'application des codes, normes et règlements relatifs au vieillissement des centrales nucléaires.

Pourquoi ce glossaire ?

L'Agence pour l'énergie nucléaire (AEN) a publié ce glossaire, en co-opération avec la Commission des Communautés européennes (CCE) et l'Agence internationale de l'énergie atomique (AIEA), pour servir d'outil pratique de référence destiné à faciliter et encourager l'utilisation, par le plus grand nombre, d'une terminologie commune du vieillissement. L'objectif recherché est de fournir au personnel des centrales (ainsi qu'à d'autres personnes s'occupant du vieillissement des installations) un ensemble commun de termes ayant la même signification pour l'ensemble de l'industrie.

Dans chacune des sections de terminologie, les termes sont classés par ordre alphabétique avec des chiffres de référence. Ces chiffres sont répétés dans la section de langue anglaise permettant ainsi d'établir des liens de correspondance entre les langues. Les tirets dans les colonnes des nombres de référence indiquent qu'il n'y a pas de termes équivalents dans certaines langues. Ces mêmes tirets n'apparaissent que dans les colonnes des synonymes.

Dans une première partie du glossaire sont examinés tous les termes regroupés en six catégories. Puis vient une liste alphabétique des termes avec leurs définitions et quelques exemples. Les dernières pages du glossaire contiennent des schémas et une liste d'idées clés apportant des éclaircissements sur la terminologie.

Remerciements

Ce glossaire suit de très près une publication faite par l'EPRI (Institut de recherches sur l'énergie électrique) (BR-101747) que nous tenons à remercier pour son soutien et ses conseils dans l'édition de cette version internationale. Le glossaire de l'EPRI a été développé grâce à des contributions provenant de différentes compagnies d'électricité, de l'Institut de l'énergie nucléaire (NEI), de la Commission de la réglementation nucléaire (USNRC) et de laboratoires américains.

Expressions usuelles, par catégorie, ayant trait au vieillissement (synonymes non inclus)

DÉGRADATION	
CAUSES DE DÉGRADATION	**DÉGRADATION ET/OU VIEILLISSEMENT**
condition • conditions de service • conditions avant mise en service • conditions ambiantes • conditions fonctionnelles • conditions d'exploitation • conditions normales • conditions induites par erreur • événement de référence • conditions d'événements de référence • conditions nominales facteur de contrainte • facteur de contrainte normal • facteur de contrainte induit par erreur • facteur de contrainte dû aux événements de référence	caractéristique condition • état dégradé vieillissement • vieillissement naturel • vieillissement prématuré • vieillissement normal • vieillissement artificiel • vieillissement accéléré • établissement des conditions propres à l'âge mécanisme de vieillissement effets du vieillissement • effets combinés • effets simultanés • effets synergiques dégradation • dégradation par vieillissement • dégradation par vieillissement normal • dégradation par vieillissement induit par erreur évaluation du vieillissement
CYCLE DE VIE	
DURÉE DE VIE	**DÉFAILLANCE**
âge durée en service durée de vie • durée de vie en place • durée de vie en service • durée de vie restante • durée de vie nominale • durée de vie nominale restante • durée de vie homologuée réforme	défaillance • défaillance dégradée • défaillance complète • défaillance aléatoire • défaillances imputables à des causes communes • défaillances de mode commun • usure cause de défaillance • cause profonde • mécanisme de défaillance • mode de défaillance analyse de la défaillance • évaluation de la défaillance • analyse des modes de défaillance et de leurs effets • établissement de l'évolution des défaillances moyenne des temps de bon fonctionnement
GESTION DU VIEILLISSEMENT	
MAINTENANCE	**ÉVALUATION DE L'ÉTAT**
gestion du vieillissement gestion de la durée de vie maintenance • maintenance préventive • maintenance périodique • maintenance programmée • maintenance corrective réparation rénovation révision remplacement entretien essais après maintenance remise en conformité	maintenance prévisionnelle inspection en service essai en service surveillance prescriptions en matière de surveillance suivi de l'état indicateur d'état indicateur fonctionnel essais diagnostic critère de réception

		anglais	allemand	espagnol	russe
1	**âge :** temps compris entre la fabrication d'un SSC[1] et un moment déterminé	3	4	38	5
2	**analyse de la cause profonde :** synonyme *d'analyse de la défaillance*	97	20	5	2
3	**analyse de la défaillance :** procédé systématique de détermination et de documentation des mécanismes et des causes superficielles et profondes de la défaillance d'un SSC.	43	20	6	3
4	**analyse des modes de défaillance et de leurs effets :** processus systématique visant à déterminer et à documenter des modes de défaillance potentiels et leurs effets sur les SSC	48	42	7	1
5	**caractéristique :** propriété ou attribut d'un SSC (tel que la forme, la dimension, le poids, l'indicateur d'état, l'indicateur fonctionnel, la performance, ainsi que la propriété mécanique, chimique ou électrique)	14	38	9	91
6	**cause de défaillance :** circonstances au cours de la conception, de la fabrication, des essais ou de l'utilisation qui ont conduit à la défaillance	44	23	10	54
7	**cause profonde :** raison(s) fondamentale(s) d'une condition observée d'un SSC qui, si elle(s) est (sont) corrigée(s), empêche(nt) la récurrence de cette condition	96	39	11	25
8	**condition :** (1) état physique environnant ou influences susceptibles d'affecter un SSC ; (2) état ou niveau des caractéristiques d'un SSC qui peut se répercuter sur son aptitude à effectuer une fonction nominale	19, 20	31	12	75
9	**conditions ambiantes :** états physiques du milieu environnant d'un SSC • *exemples* : température, rayonnement et humidité dans l'enceinte de confinement dans les conditions normales de fonctionnement ou pendant les accidents	38	85	15	78
10	**conditions avant mise en service :** états ou influences physiques effectifs s'exerçant sur un SSC préalablement au fonctionnement initial (fabrication, stockage, transport, installation et essais préliminaires)	81	33, 92	16	52

---

1. SSC = Système, structure ou composant.

		anglais	allemand	espagnol	russe
11	**conditions d'événements de référence :** conditions de service imputables à des événements de référence	30	32	23	77
12	**conditions d'exploitation :** conditions de service, y compris les conditions normales et induites par erreur, avant le début d'un accident de dimensionnement ou d'un séisme	70	35	19	95
13	**conditions de dimensionnement :** synonyme de *conditions nominales*	34	25	22	–
14	**conditions de service :** états ou influences physiques effectifs au cours de la durée de vie utile d'un SSC, y compris les conditions d'exploitation (normales et induites par erreur), les conditions d'événement de référence et les conditions après événement de référence	98	40, 52	21	60
15	**conditions de service en cours d'exploitation :** synonyme de *conditions d'exploitation*	71	35	–	–
16	**conditions fonctionnelles :** influences s'exerçant sur un SSC du fait de l'exécution de fonctions pour lesquelles il est conçu (fonctionnement d'un système ou d'un composant et chargement d'une structure) • *exemples :* – pour un clapet anti-retour, cycle de fonctionnement et battement ; – pour une soupape de décharge de la cuve du réacteur pression primaire, vitesses d'écoulement élevées et augmentation de la température primaire	50	43	24	90
17	**conditions induites par erreur :** conditions avant mise en service ou de service néfastes imputables à des erreurs de conception, de fabrication, d'installation, d'exploitation ou de maintenance	40	14, 46	25	76
18	**conditions nominales :** conditions de service spécifiées, qui sont utilisées pour établir les spécifications d'un SSC (comportant généralement une marge de sécurité au-delà des conditions de service escomptées)	32	25	18	57
19	**conditions normales :** conditions d'exploitation d'un SSC convenablement conçu, fabriqué, installé, exploité et entretenu, à l'exclusion des conditions d'événements de référence	67	61	26	34

		anglais	allemand	espagnol	russe
20	**conditions normales d'exploitation :** synonyme de *conditions normales*	68	61	20	35
21	**conditions opérationnelles :** synonyme de *conditions fonctionnelles*	72	43	27	–
22	**critère d'acceptabilité :** limite spécifiée d'un indicateur fonctionnel ou d'état utilisée pour évaluer l'aptitude d'un SSC à remplir la fonction pour laquelle il est conçu	2	3	29	26
23	**défaillance :** inaptitude ou interruption de l'aptitude d'un SSC à fonctionner dans les limites des critères d'acceptabilité	42	15, 86	54	39
24	**défaillance aléatoire :** toute défaillance dont la cause ou le mécanisme, ou les deux, en rendent imprévisible le moment de survenue	84	98	55	64
25	**défaillance complète :** défaillance dans laquelle il y a perte complète de fonction	18	79	56	51
26	**défaillances de mode commun :** deux ou plusieurs défaillances se produisant de la même manière ou selon le même mode, qui sont dues à une seule cause	17	19, 77	58	14
27	**défaillances imputables à des causes communes :** deux ou plusieurs défaillances dues à une seule cause	16	16, 77	57	41
28	**défaut :** synonyme de *défaillance*	62	15, 86	–	32
29	**dégradation :** détérioration immédiate ou graduelle des caractéristiques d'un SSC qui pourrait altérer son aptitude à fonctionner dans les limites des critères d'acceptabilité	25	2, 88	30	80
30	**dégradation liée à l'âge :** synonyme de *dégradation par vieillissement*	11	7, 8	34	–
31	**dégradation par vieillissement :** effets du vieillissement qui pourraient altérer l'aptitude d'un SSC à fonctionner dans les limites des critères d'acceptabilité	7	7, 8	31	81

- *exemples :* réduction du diamètre d'un arbre due à l'usure, perte de résistance des matériaux induite par la fatigue ou le vieillissement thermique, gonflement des matériaux d'étanchéité et perte de la rigidité diélectrique ou fissuration de l'isolant

		anglais	allemand	espagnol	russe
32	**dégradation par vieillissement induit par erreur :** dégradation par vieillissement due à des conditions induites par erreur	39	13	32	83
33	**dégradation par vieillissement normal :** dégradation par vieillissement due à des conditions normales	66	63	33	82
34	**détérioration :** synonyme de *dégradation*	35	2, 88	36	–
35	**diagnostic :** examen et évaluation des données afin de déterminer soit l'état d'un SSC, soit les causes de cet état	36	37	37	9
36	**durée de vie :** période qui va de la fabrication d'un SSC à sa réforme	56	53	90	13
37	**durée de vie en place :** période qui va de l'installation d'un SSC à sa réforme	55	47	94	68
38	**durée de vie en service :** période qui va de la mise en exploitation initiale d'un SSC à sa réforme	99	41, 65, 75	92	62
39	**durée de vie en service restante :** synonyme de *durée de vie restante*	89	72	93	–
40	**durée de vie homologuée :** période pendant laquelle il a été démontré par des essais, l'analyse ou l'expérience qu'un SSC est capable de fonctionner dans les limites des critères de réception en présence de conditions d'exploitation spécifiées tout en conservant son aptitude à remplir ses fonctions de sûreté en cas d'accident de référence ou de séisme	83	45	76	79
41	**durée de vie nominale :** période pendant laquelle il est prévu qu'un SSC fonctionnera dans les limites des critères d'acceptabilité	33	27	91	58
42	**durée de vie nominale restante :** période qui va d'un moment déterminé jusqu'à la réforme programmée d'un SSC	87	71	96	37
43	**durée de vie résiduelle :** synonyme de *durée de vie restante*	93	72	–	–
44	**durée de vie restante :** période qui va d'un moment déterminé jusqu'à la réforme d'un SSC	88	72	95	38

		anglais	allemand	espagnol	russe
45	**durée de vie utile :** synonyme de *durée de vie en service*	110	41, 65, 75	97	–
46	**durée de vie utile restante :** synonyme de *durée de vie restante*	90	72	98	–
47	**durée en service :** temps compris entre la mise en exploitation initiale d'un SSC et un moment déterminé	109	36	87	46
48	**effets combinés :** modifications nettes des caractéristiques d'un SSC imputables à deux ou plusieurs facteurs de contrainte	15	50	39	65
49	**effets du vieillissement :** modifications nettes des caractéristiques d'un SSC qui se produisent avec le temps ou l'utilisation et qui sont dues aux mécanismes de vieillissement • *exemples :* – effet négatif : voir *dégradation par vieillissement* – effet positif : augmentation de la résistance du béton due au durcissement; moindre vibration par suite de l'usure des machines tournantes	8	6	40	96
50	**effets simultanés :** effets combinés dus à des facteurs de contrainte agissant simultanément	101	74	42	36
51	**effets synergiques :** part des modifications des caractéristiques d'un SSC imputable exclusivement à l'interaction de facteurs de contrainte agissant simultanément, par opposition aux modifications résultant de la superposition de l'action de chaque facteur de contrainte agissant indépendamment	107	76	41	63
52	**entretien :** mesures courantes (y compris opérations de nettoyage, d'ajustement, d'étalonnage et de remplacement des éléments consommables) qui maintiennent ou prolongent la durée de vie utile d'un SSC	100	94	82	70
53	**essai en service :** essai visant à déterminer la disponibilité opérationnelle d'un composant ou d'un système	54	96	45	93
54	**essai :** observation ou mesure d'indicateurs d'état dans des conditions contrôlées afin de vérifier qu'un SSC est conforme aux critères d'acceptabilité	108	68, 78	43	19

		anglais	allemand	espagnol	russe
55	**essai après maintenance :** essai effectué après maintenance en vue de vérifier que celle-ci a été correctement exécutée et que le SSC peut fonctionner dans les limites des critères d'acceptabilité	77	69	44	20
56	**essai de surveillance :** synonyme de *surveillance, prescriptions en matière de surveillance*, et *essai* (utilisé seulement lorsque des connotations réglementaires et juridiques spécifiques sont requises)	106	–	–	30
57	**établissement de l'évolution de l'état :** synonyme de *suivi de l'état*	23	101	–	56
58	**établissement de l'évolution des défaillances :** enregistrement, analyse et extrapolation des défaillances en cours d'exploitation d'un SSC en fonction d'un paramètre indépendant	49	80	86	55
59	**établissement des conditions :** synonyme *d'établissement des conditions propres à l'âge*	78	89	–	–
60	**établissement des conditions propres à l'âge :** simulation des effets du vieillissement naturel d'un SSC par l'application de toute combinaison d'un vieillissement artificiel et naturel	4	89	52	29
61	**état dégradé :** état d'acceptabilité d'un SSC non défaillant à la limite de l'acceptabilité et qui pourrait conduire à la décision d'exécuter une maintenance programmée	26	1	14	84
62	**évaluation à des fins de diagnostic :** synonyme de *diagnostic*	37	37	–	10
63	**évaluation de la défaillance :** conclusions tirées d'une analyse de la défaillance	45	21	53	43
64	**évaluation de la durée de vie :** synonyme de *évaluation du vieillissement*	57	54	–	42
65	**évaluation du vieillissement :** évaluation d'informations appropriées permettant de déterminer les effets du vieillissement sur l'aptitude actuelle et future des SSC à fonctionner dans les limites des critères d'acceptabilité	6	9	89	44

		anglais	allemand	espagnol	russe
66	**événement de référence :** l'un quelconque des événements spécifiés dans l'analyse de sûreté de la centrale qui sont utilisés pour s'assurer que les fonctions liées à la sûreté des SSC sont exécutées de façon acceptable; ces événements couvrent les transitoires prévus, les accidents de dimensionnement, les agressions d'origine externe et les phénomènes naturels	29	26	84	21
67	**facteur de contrainte :** agent ou sollicitation qui découle des conditions avant mise en service et de service et qui peut entraîner une dégradation immédiate ou par vieillissement d'un SSC • *exemples* : chaleur, rayonnement, humidité, vapeur, produits chimiques, pression, vibration, mouvement sismique, cycle de fonctionnement électrique et cycle de fonctionnement mécanique	102	30	1	85
68	**facteur de contrainte dû aux événements de référence :** facteur de contrainte qui découle des événements de référence et qui peut entraîner une dégradation immédiate, ou une dégradation par vieillissement dépassant celle qu'entraînent des facteurs de contrainte normaux	31	24	4	87
69	**facteur de contrainte induit par erreur :** facteur de contrainte qui découle de conditions induites par erreur et peut entraîner une dégradation immédiate, ou une dégradation par vieillissement dépassant celle qu'entraînent des facteurs de contrainte normaux	41	12	2	86
70	**facteur de contrainte normal :** facteur de contrainte qui découle de conditions normales et peut entraîner des mécanismes de vieillissement et des effets de vieillissement dans un SSC	69	64	3	88
71	**fonctionnement dégradé :** défaillance dans laquelle un indicateur fonctionnel ne répond pas au critère d'acceptabilité, mais où la fonction pour laquelle il est conçu n'est pas entièrement perdue	27	18	59	40

		anglais	allemand	espagnol	russe
72	**gestion de la durée de vie :** intégration de la gestion du vieillissement et de la planification économique afin : (1) d'optimiser l'exploitation, la maintenance et la durée de vie utile des SSC ; (2) de maintenir un niveau acceptable de performances techniques et de sûreté ; et (3) de maximiser le rendement de l'investissement pendant la durée de vie utile de la centrale	59	55	60	72
73	**gestion du cycle de vie :** synonyme de *gestion de la durée de vie*	58	56	–	–
74	**gestion du vieillissement :** mesures d'ordre technique, d'exploitation et de maintenance visant à maintenir, dans des limites acceptables la dégradation par vieillissement et l'usure des SSC • *exemples* d'activités d'ingénierie : conception, qualification et analyse des défaillances • *exemples* d'actions de conduite : surveillance, exécution des procédures de conduite dans des limites précisées et réalisation de mesures dans l'environnement	9	10	61	73
75	**indicateur d'état :** caractéristique qui peut être observée, mesurée ou dont l'évolution peut être établie afin d'en déduire ou d'indiquer directement l'aptitude actuelle et future d'un SSC à fonctionner dans les limites des critères d'acceptabilité	21	100	62	50
76	**indicateur de performance :** synonyme *d'indicateur fonctionnel*	74	44	63	49
77	**indicateur fonctionnel :** indicateur d'état qui constitue une indication directe de l'aptitude actuelle d'un SSC à fonctionner dans les limites des critères d'acceptabilité	51	44	64	89
78	**inspection en service :** examen ou contrôle de l'intégrité d'un SCC pendant l'exploitation ou l'arrêt de la centrale	52	73, 97	66	92
79	**maintenance :** ensemble des actions directes ou indirects (de rapport) qui permettent de déceler, d'éviter ou d'atténuer la dégradation d'un SSC en fonctionnement ; ou de rétablir à un niveau acceptable l'aptitude d'un SSC défaillant à remplir les fonctions nominales	61	48, 95	67	71

		anglais	allemand	espagnol	russe
80	**maintenance corrective :** mesures qui rétablissent, par réparation, révision ou remplacement, l'aptitude d'un SSC défaillant à fonctionner dans les limites des critères d'acceptabilité	24	49	68	18
81	**maintenance périodique :** forme de maintenance préventive consistant à assurer l'entretien, le remplacement de pièces, la surveillance ou l'exécution d'essais à intervalles de temps prédéterminés, à intervalles de temps de durée de fonctionnement ou de nombre de cycles prédéterminés	75	66	69	45
82	**maintenance préventive :** mesures qui détectent, préviennent ou atténuent la dégradation d'un SSC fonctionnel de manière à maintenir ou prolonger sa durée de vie utile en faisant en sorte que la dégradation et les défaillances demeurent à un niveau acceptable. La maintenance préventive est de trois types : périodique, prévisionnelle et programmée	82	91	72	47
83	**maintenance prévisionnelle :** forme de maintenance préventive exécutée de façon continue ou à intervalles régis par l'état observé, afin de suivre, diagnostiquer ou déterminer l'évolution d'indicateurs fonctionnels ou d'état d'un SSC. Les résultats fournissent des indications quant à l'aptitude actuelle et future du SSC à fonctionner ou à la nature et au calendrier de la maintenance programmée	79	90	71	59
84	**maintenance programmée :** forme de maintenance préventive consistant en une rénovation, une révision ou un remplacement qui sont planifiés et exécutés avant la défaillance d'un SSC	76	67	70	48
85	**mécanisme de défaillance :** processus physique qui entraîne une défaillance • *exemples* : fissuration de l'isolant d'un câble fragilisé (provoquée par le vieillissement) ; objet entravant l'écoulement (sans rapport avec le vieillissement)	46	22	74	27
86	**mécanisme de vieillissement :** processus spécifique qui modifie graduellement les caractéristiques d'un SSC avec le temps ou l'utilisation • *exemples* : durcissement, usure, fatigue, fluage, érosion, salissure microbiologique, corrosion, fragilisation et réactions chimiques ou biologiques	10	11	73	28

		anglais	allemand	espagnol	russe
87	**mode de défaillance :** manière dont un SSC devient défaillant ou état dans lequel il subit une défaillance	47	87	75	4
	• *exemples* : blocage en position ouverte (vanne), court-circuit à la masse (câble), grippage d'un palier (moteur), fuite (vanne, cuve ou enceinte de confinement), obstruction (tuyauterie ou vanne), non émission d'un signal qui fait descendre les barres de commande (système de protection du réacteur) et fissure ou rupture (structure)				
88	**moyenne des temps de bon fonctionnement :** moyenne arithmétique des temps de fonctionnement séparant les défaillances d'une pièce	63	57	88	67
89	**prescriptions en matière de surveillance :** essai, étalonnage ou inspection visant à garantir que la qualité requise des systèmes et composants est maintenue, que l'exploitation de l'installation se situera à l'intérieur des limites de sûreté, et que les conditions limitatives de fonctionnement seront respectées (n'utiliser que lorsque des connotations réglementaires et juridiques spécifiques sont requises)	105	84	80	31
90	**réforme :** retrait définitif du service d'un SSC	94	28	81	8
91	**remise en conformité :** correction d'une fabrication, d'une installation ou d'une maintenance qui avait été exécutée de façon inadéquate	95	58	77	11
92	**remise en état :** synonyme de *révision*	85	81	–	–
93	**remplacement :** retrait d'un SSC non dégradé, dégradé ou défaillant, ou d'un élément de celui-ci, et installation d'un autre, à sa place, qui est capable de fonctionner dans les limites des critères initiaux d'acceptabilité	92	29	85	15
94	**rénovation :** mesures programmées pour améliorer la condition d'un SSC non défaillant	86	59	78	6
95	**réparation :** mesures en vue de ramener un SSC défaillant dans un état acceptable	91	70	79	61
96	**révision :** réparation importante, rénovation, ou les deux	73	81	83	22

		anglais	allemand	espagnol	russe
97	**suivi de l'état :** observation, mesure ou détermination de l'évolution des indicateurs d'état ou fonctionnels en fonction d'un paramètre indépendant (habituellement le temps ou le nombre de cycles) afin d'indiquer l'aptitude actuelle et future d'un SSC à fonctionner dans les limites des critères d'acceptabilité	22	101	28	24
98	**surveillance en exploitation :** observation ou mesure d'indicateurs d'état ou fonctionnels pour vérifier qu'un SSC est actuellement capable de fonctionner dans les limites des critères d'acceptabilité	103	82	65	94
99	**usure :** défaillance due à un mécanisme de vieillissement	111	17	35	16
100	**vie :** synonyme de *durée de vie*	60	53	–	–
101	**vieillissement accéléré :** vieillissement artificiel consistant à simuler dans un bref laps de temps, les effets du vieillissement naturel dû aux conditions de service sur une période plus longue (voir aussi vieillissement prématuré)	1	34	47	74
102	**vieillissement artificiel :** simulation des effets du vieillissement naturel sur des SSC par application de facteurs de contrainte représentant les conditions avant mise en service et de service ; les conditions simulées peuvent être différentes des conditions réelles en intensité, durée et mode d'application	12	51	48	17
103	**vieillissement naturel :** vieillissement d'un SSC qui intervient dans des conditions avant mise en service et de service, y compris des conditions induites par erreur	64	60	49	12
104	**vieillissement normal :** vieillissement naturel imputable à des conditions avant mise en service et de service exemptes d'erreur	65	62	50	33
105	**vieillissement prématuré :** vieillissement qui intervient plus tôt que prévu	80	93	51	53
106	**vieillissement :** processus général par lequel les caractéristiques d'un SSC se modifient graduellement avec le temps ou l'utilisation	5	5	46	69

IDÉES CLÉS DE LA TERMINOLOGIE COMMUNE DU VIEILLISSEMENT

Causes de dégradation

- Les **conditions de service** sont toutes les conditions réelles qui ont un impact sur le SSC. Il s'agit des **conditions de fonctionnement** (y compris les **conditions normales** et les **conditions induites par erreur** ainsi que les transitoires envisagés) et des conditions accidentelles.

- Les **conditions nominales** sont les conditions hypothétiques incorporant généralement par définition une marge de sécurité au-delà des conditions de service réelles escomptées.

Durée de vie

- La **durée de vie utile** est la période réelle pendant laquelle un SSC fournit un service utile. Cette dernière peut être différente de la durée de vie utile escomptée, à savoir la **durée de vie nominale**.

- L'**âge** d'un SSC (mesuré à partir du moment de sa fabrication) peut ne pas correspondre à sa **durée de service** (mesurée à partir de son fonctionnement initial).

Dégradation/vieillissement

- La **dégradation** est progressive (vieillissement) ou immédiate (non causée par le vieillissement).

- La **dégradation par vieillissement** est engendrée par les **conditions d'exploitation**, y compris les **conditions ambiantes**, comme la température et le rayonnement, ainsi que les **conditions fonctionnelles**, comme les déplacements des pièces les unes par rapport aux autres. Les conditions d'exploitation engendrent des **facteurs de contrainte normaux** ou des **facteurs de contraintes induits par erreur**.

- Les **événements de référence** comprennent les transitoires prévus durant le fonctionnement de la centrale (qui peuvent contribuer au **vieillissement**) ainsi que les accidents de dimensionnement et les séismes (qui sont à l'origine d'une **dégradation** immédiate qui n'est pas liée au vieillissement).

- Un **état dégradé** est à la limite de l'acceptabilité mais n'est pas une **défaillance**. Dans le cas de **défaillances par dégradation**, les **critères d'acceptabilité** pour les conditions et/ou les performances ne sont pas satisfaites, mais la fonction est partiellement exécutée. En cas de **défaillance complète**, il y a perte de la fonction.

Défaillance

- Une **défaillance** est habituellement engendrée par une séquence de causes et non par une cause unique. L'**usure** est une défaillance dont la dernière cause est un **mécanisme de vieillissement**. La **cause profonde** peut ne pas être ce mécanisme de vieillissement.

- Le **vieillissement prématuré** peut engendrer une **défaillance** d'un SSC en service. Le terme de **vieillissement accéléré** doit être réservé au **vieillissement artificiel** habituellement obtenu dans un laboratoire.

- Les **analyses de défaillance** permettent de déterminer les causes de défaillance, le **mécanisme de défaillance** et le **mode de défaillance**. Chacun de ces termes a un sens différent. La **dégradation par vieillissement** et les **défaillances induites par erreur** n'ont pas pour **cause profonde** le **vieillissement** mais une erreur humaine.

Maintenance/évaluation de l'état

- La **maintenance** est un terme général qui recouvre la **maintenance corrective** et la **maintenance préventive**. La **maintenance** peut être effectuée par le personnel de maintenance, le personnel d'exploitation ou le personnel technique. La **maintenance préventive** comprend la **maintenance prévisionnelle**, comme la **surveillance**, les **essais** et le **suivi de l'état.**

- Les **réparations** ne sont effectuées que sur un SSC défaillant ; une **rénovation** n'est réalisée que sur un SSC non défaillant. Une **révision** est une réparation importante et/ou une rénovation.

Figure 1.   **Relations entre les catégories d'expressions**

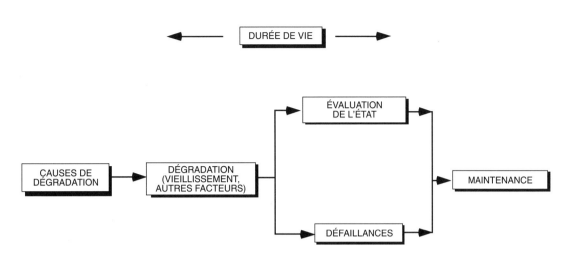

Note :   L'évaluation de l'état constitue en fait la partie prévisionnelle de la maintenance.

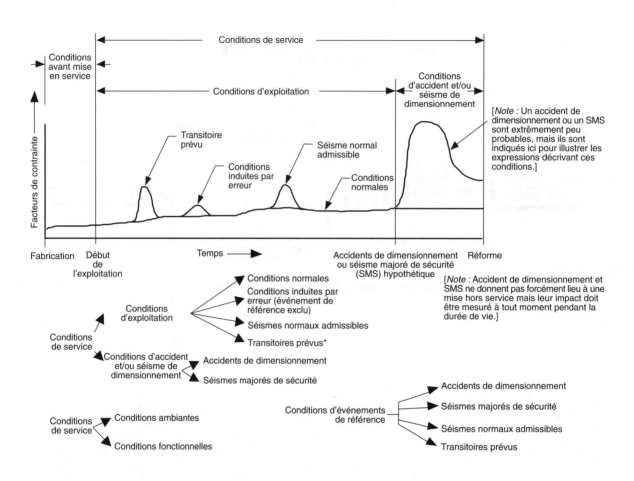

Figure 2. **Relations entre les types de conditions de service**

* Il convient de noter que les transitoires prévus, bien qu'ils soient retenus dans le dimensionnement, peuvent être induits par erreur. Les conditions nominales couvrent toutes les conditions de service, à l'exception des conditions non liées à des événements de référence induites par erreur.

Figure 3. **Relations entre les expressions relatives au vieillissement**

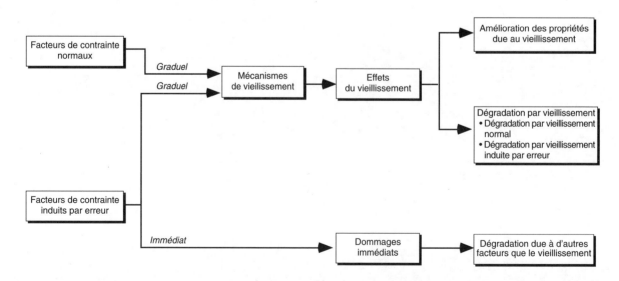

48

Figure 4. **Relations entre les expressions décrivant des événements de la durée de vie réelle par rapport à la séquence des événements dans le temps**

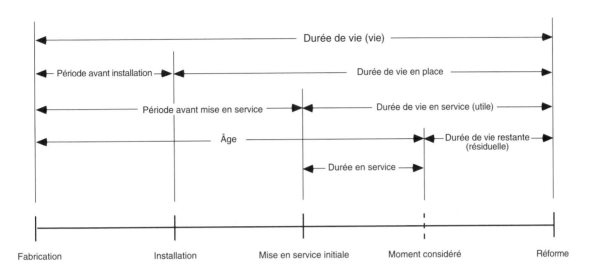

Note : Ces termes définissent les étapes effectives de l'évolution passée ou à venir d'un système, d'une structure ou d'un composant. Les expressions synonymes sont indiquées entre parenthèses.

Figure 5. **Relations entre les expressions décrivant les prévisions de durée de vie nominale**

A. Prévision initiale concernant la durée de vie nominale

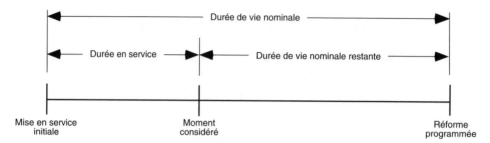

B. Prolongation possible de la durée de vie nominale par suite d'une révision (discrétionnaire) de la durée de vie nominale prévue

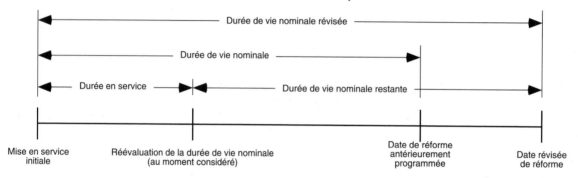

C. Abrégement possible de la durée de vie normale par suite d'une révision (discrétionnaire) de la durée de vie nominale prévue

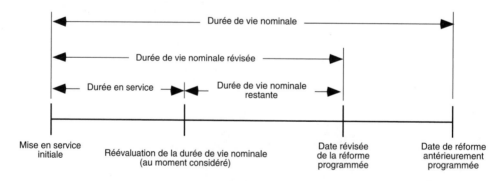

Figure 6. **Relations entre les expressions relatives aux défaillances**

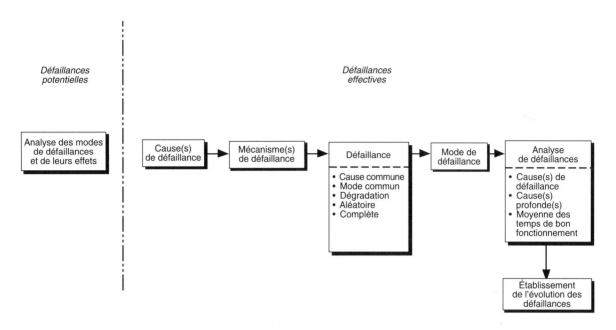

Figure 7. **Exemples de causes, mécanismes et modes**

Causes profondes et usure

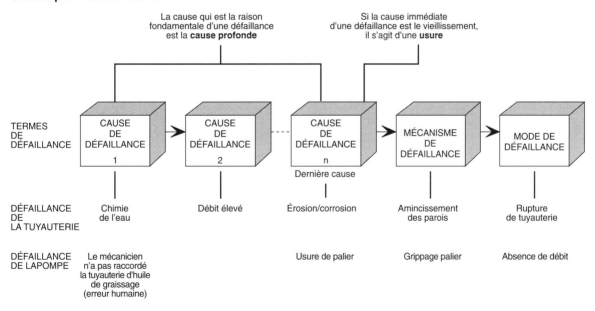

Figure 8. **Termes décrivant la maintenance**

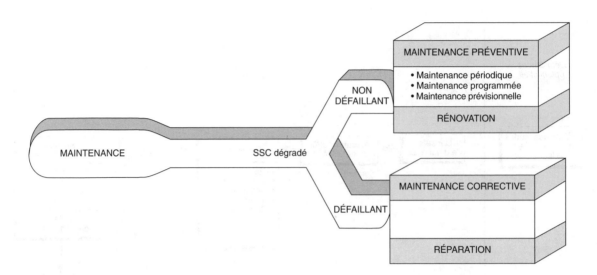

Figure 9. **Relations entre les termes de maintenance**

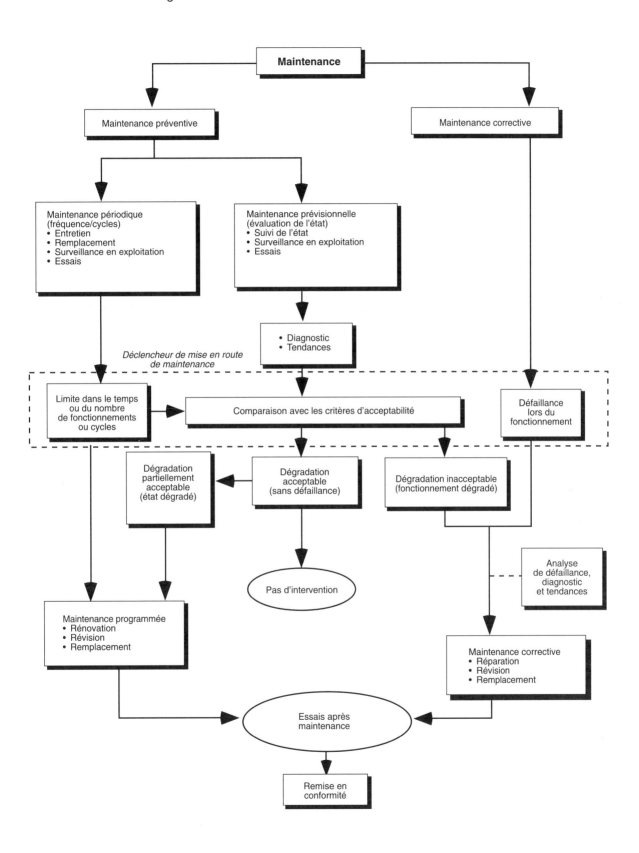

GLOSSAR DER ALTERUNG KERNKRAFTWERKEN

Ein nützliches Glossar für verstehen und handhaben

der Alterung von Systemen, Strukturen

und Komponenten in Kernkraftwerken

WOZU DIE BEGRIFFE-SAMMLUNG ZUR ALTERUNG?

Im gleichen Maße wie die Nutzungsdauer der laufenden Kernkraftwerke zunimmt, verdienen mögliche Fehlinterpretationen der Alterungsprozesse von Systemen, Strukturen oder Komponenten (SSK) größere Aufmerksamkeit. Die Begriffe-Sammlung zur Alterung wurde zusammengestellt, um das Verständnis für die Alterungsphänomene zu heben, die Meldung relevanter Anlagenausfalldaten zu vereinfachen und die einheitliche Interpretation von Vorschriften zu unterstützen.

Die Begriffe-Sammlung könnte beim Lebensdauer- und Alterungsmanagement behilflich sein. Lebensdauermanagement kann Betriebs- und Instandhaltungskosten minimieren und die Option zur Lebensdauerverlängerung einer Anlage von 40 auf 60 Jahre stützen. Weit wichtiger noch ist die Bedeutung eines effektiven Alterungsmanagements zur Erhaltung eines adäquaten Sicherheitsniveaus der Anlage.

Die Einsicht in die Bedeutung einer eindeutigen Kommunikation auf diesem Gebiet hat Vertreter verschiedener Unternehmen veranlaßt, eine einheitliche Begriffe-Sammlung zum Thema Alterung zu erarbeiten.

Angesichts der Vorteile empfiehlt sich die Verwendung der Begriffe-Sammlung zur Alterung für technische Dokumente, Berichte zu Ausfällen, Forschungsberichte, künftige Regelungen und für andere Unterlagen, die sich mit Alterung befassen. Die Unterlagen sollten auf die Verwendung der Begriffe-Sammlung zur Alterung hinweisen, Ausnahmen sollten angegeben werden. Solche Ausnahmen sind Fälle, in denen die Verfasser die Definitionen bestehender Vorschriften vorziehen.

Hauptvorteile einer Verwendung der Begriffe-Sammlung zur Alterung sind folgende:

- verbesserte Übermittlung und Interpretation von Anlagendaten zu Verschleiß und Ausfall von SSK, einschließlich der zutreffenden Ursachenfeststellung;

- verbesserte Interpretation von und Übereinstimmung mit Vorschriften zur Alterung in Kernkraftwerken.

Wozu deise Ausgabe?

NEA (OECD-Kernenergie-Agentur) hat in Zusammenarbeit mit CEC (Kommission der Europäischen Gemeinschaften) und IAEA (Internationale Atomenergie-Organisation) diese Ausgabe veröffenlichtlegt, als handliche Hilfe, um zu einer breiten Anwendung der Begriffe-Sammlung zur Alterung zu ermutigen. Das Ziel ist, dem Anlagenpersonal (und anderen, die sich mit Alterung beschäftigen) ein Sammlung von gebräulichen Begriffen an die Hand zu geben mit einheitlicher, industrieweiter Bedeutung, um die Diskussion zwischen Experten verschiedener Länder zu vereinfachen.

In jeder Sprache sind die Begriffe alphabetisch geordnet und mit Nummern versehen. Diese Nummern wiederholen sich im englisch-sprachigen Abschnitt und erlauben so Querverweise zwischen allen Sprachen. Trennungsstriche in Spalten mitteilen, dass es keine Gegenstücke in gewissen Sprächen geben. Trennungsstriche erscheinen nur in Spalten mit synonymen Ausdrücken.

In jeder Sprache beginnt die Sammlung mit einer Übersicht aller Begriffe, eingeteilt in sechs Kategorien. Dem folgt eine alphabetische Liste der Begriffe, der Definitionen und einiger Beispiele. Die letzten Seiten von jedem Abschnitt enthalten Diagramme und eine Zusammenstellung der Basisideen, um die relativen Begriffsbedeutungen zu erhellen.

Danksagung

Diese Begriffe-Sammlung lehnt sich eng an eine Veröffentlichung durch EPRI (Stromforschungsinstitut) (BR-101747) an, dem wir für die Unterstützung und die Ratschläge in Zuge der Erarbeitung dieser internationalen Version danken. Die EPRI Begriffe-Sammlung stützt sich auf Beitrage von mehreren Amerikanischen Elektrizitätsversorgungsunternehmen, dem Kernenergie-Institut (NEI), und Amerikanischen Laboratorien.

Allgemeine Alterungsbegriffe, sortiert nach Kategorie (ohne Synonyme)

VERSCHLEIß	
URSACHEN FÜR VERSCHLEIß	**VERSCHLEIß/ALTERUNG**
Bedingung • Einsatzbedingungen • Voreinsatzbedingungen • Umgebungsbedingungen • Funktionsbedingungen • bestimmungsgemäßer Betrieb • Normalbetrieb • anomaler Betrieb, gestörter Betrieb • Auslegungsereignis • Bedingungen durch Auslegungsereignisse • Auslegungsbedingungen Beanspruchung • normale Beanspruchung • anomale Beanspruchung • Auslegungsbeanspruchung	Eigenschaft Zustand • abgenutzter Zustand Alterung • natürliche Alterung • vorzeitige Alterung • normale Alterung • künstliche Alterung • beschleunigte Alterung • Voralterung Alterungsmechanismus Alterungsauswirkungen • kombinierte Wirkungen • Simultanwirkungen • synergistische Wirkungen Verschleiß • alterungsbedingte Abnutzung • normale alterungsbedingte Abnutzung • anomale alterungsbedingter Verschleiß Alterungsbeurteilung
LEBENSZYKLUS	
LEBENSDAUER	**AUSFALL**
Alter Betriebsdauer Lebensdauer • installierte Lebensdauer • Nutzungsdauer • Restlebensdauer • Auslegungslebensdauer • Restauslegungslebensdauer • geplante Standzeit Außerbetriebnahme	Ausfall • Ausfall durch Verschleiß • Totalausfall • Zufallsausfall • Ausfall aus gemeinsamer Ursache • Ausfall gleicher Art • systematischer Fehler • Ausfall durch Alterung Ausfallursache • eigentlicher Grund • Ausfallmechanismus • Versagensart Ausfallanalyse • Ausfallbewertung • Fehlereffektanalyse • Trendbildung der Ausfälle mittlere Ausfallzeit
ALTERUNGSMANAGEMENT	
INSTANDHALTUNG	**ZUSTANDSBEWERTUNG**
Alterungsmanagement Lebensdauermanagement Instandhaltung • vorbeugende Instandhaltung • periodische Instandhaltung • planmäßige Instandhaltung • Instandsetzung Reparatur Nachrüstung Überholung Austausch Wartung Prüfung nach Instandhaltungsarbeiten Nachbesserung	vorauseilende Instandhaltung wiederkehrende Prüfung Wiederholungsprüfung Überwachung Überwachungsanforderung Zustandsüberwachung Zustandsindikator Funktionsindikator Prüfung Diagnose Akzeptanzkriterium

		Englisch	Französisch	Spanisch	Russisch
1	**abgenutzter Zustand:** gerade noch akzeptierbarer Zustand nicht ausgefallener SSK[1], der zu einer planmäßigen Instandhaltung führen könnte	26	61	14	84
2	**Abnutzung:** Synonym für *Verschleiß*	25, 35	29	30	80
3	**Akzeptanzkriterium:** spezifizierte Grenzen eines Funktions- oder Zustandsindikators, der zur Beurteilung der Fähigkeit der SSK dient, ihre Auslegungsfunktion zu erfüllen	2	22	29	26
4	**Alter:** Zeitraum ab Herstellung der SSK bis zu einem bestimmten Zeitpunkt	3	1	38	5
5	**Alterung:** Prozeß, in dessen Verlauf sich die Eigenschaften der SSK allmählich mit fortschreitender Zeit oder Nutzung ändern	5	106	46	69
6	**Alterungsauswirkungen:** Änderungen der Eigenschaften von SSK, die durch Zeit oder Nutzung und Alterungsmechanismen hervorgerufen werden • *Beispiele*: negative Effekte - siehe **alterungsbedingte Abnutzung;** positive Effekte - Zunahme der Betonfestigkeit durch Aushärten, verringerte Vibration durch Einlaufen rotierender Maschinen	8	49	40	96
7	**alterungsbedingte Abnutzung:** Alterungswirkungen, die die Funktionsfähigkeit der SSK innerhalb der Akzeptanzkriterien beeinträchtigen könnten • *Beispiele*: Durchmesserabnahme durch Abnutzung bei rotierenden Wellen, Abnahme der Festigkeit durch Ermüdung und thermische Alterung, Schwellen von Vergußmasse, Verlust der elektrischen Durchschlagsfestigkeit oder Risse in der Isolierung	7, 11	31	31	81
8	**alterungsbedingter Verschleiß:** Synonym für *alterungsbedingte Abnutzung*	7, 11	31	31	81
9	**Alterungsbeurteilung:** Bewertung geeigneter Informationen zur Feststellung der Auswirkungen der Alterung auf die gegenwärtige und künftige Funktionsfähigkeit der SSK innerhalb der Akzeptanzkriterien	6	65	89	44

---

[1]. SSK = Systeme, Strukturen oder Komponenten

		Englisch	Französisch	Spanisch	Russisch
10	**Alterungsmanagement:** Konstruktions–, Betriebs– und Instandhaltungsmaßnahmen, um die alterungsbedingte Abnutzung und den Verschleiß der SSK innerhalb akzeptabler Grenzen zu halten • *Beispiele* für technische Maßnahmen: Auslegung, Qualifizierung und Ausfallanalyse • *Beispiele* für betriebliche Maßnahmen: Überwachung, Durchführung der Betriebsprozeduren innerhalb spezifizierter Grenzen, Überwachung der Umgebungsbedingungen	9	74	61	73
11	**Alterungsmechanismus:** spezieller Mechanismus, der allmählich die Eigenschaften der SSK mit fortschreitender Zeit oder Nutzung ändert • *Beispiele*: Aushärtung, Abnutzung, Ermüdung, Kriechen, Erosion, biologischer Bewuchs, Korrosion, Versprödung und chemische oder biologische Reaktionen	10	86	73	28
12	**anomale Beanspruchung:** Beanspruchungen, die aus dem gestörten Betrieb resultieren und zu prompter oder alterungsbedingter Abnutzung der SSK führen können und über die normale Beanspruchung hinausgehen	41	69	2	86
13	**anomaler alterungsbedingter Verschleiß:** alterungsbedingte Abnutzung, ausgelöst durch Fehler	39	32	32	83
14	**anomaler Betrieb:** ungünstige Voreinsatz- oder Einsatzbedingungen, die durch Fehler in der Auslegung, Herstellung, während der Montage, des Betriebs oder der Instandhaltung entstanden	40	17	25	76
15	**Ausfall:** Verlust der Funktionsfähigkeit der SSK innerhalb der Akzeptanzkritien	42, 62	23	54	39
16	**Ausfall aus gemeinsamer Ursache:** zwei oder mehr Ausfälle, die dieselbe Ursache haben	16	27	57	41
17	**Ausfall durch Alterung:** Ausfall, verursacht durch Alterungsmechanismen	111	99	35	16
18	**Ausfall durch Verschleiß:** Ausfall, bei dem ein Funktionsindikator nicht den Akzeptanzkriterien entspricht, die Auslegungsfunktion aber nicht vollständig versagt	27	71	59	40
19	**Ausfall gleicher Art:** zwei oder mehr Ausfälle der gleichen Art, die eine gleiche Ursache haben	17	26	58	14

		Englisch	Französisch	Spanisch	Russisch
20	**Ausfallanalyse:** systematischer Prozeß zur Ermittlung und Dokumentation von Art, Mechanismus, Ursache und Gründen für den Ausfall der SSK	43, 97	3	6	3
21	**Ausfallbewertung:** Schlußfolgerung aus der Ausfallanalyse	45	63	53	43
22	**Ausfallmechanismus:** physikalischer Prozeß, der zum Ausfall führte • *Beispiele*: Bruch versprödeter Kabelisolierung (im Zusammenhang mit Alterung), Durchflußbehinderung durch Gegenstand (kein Zusammenhang mit Alterung)	46	85	74	27
23	**Ausfallursache:** Umstände während Auslegung, Herstellung, Prüfung oder im Einsatz, die zum Ausfall führten	44	6	10	54
24	**Auslegungsbeanspruchung:** aus Auslegungsereignissen resultierender, prompter oder alterungsbedingter Verschleiß der SSK, der über das übliche Maß hinausgehen kann	31	68	4	87
25	**Auslegungsbedingungen:** Einsatzbedingungen gemäß existierender Regeln und Richtlinien, die normalerweise in technischen Spezifikationen enthalten sind (i.a. sind Sicherheitszuschläge über die erwarteten Einsatzbedingungen hinaus enthalten)	28, 32, 34	18	18	57
26	**Auslegungsereignis:** ausgewähltes Ereignis, dessen Berücksichtigung die zuverlässige Funktionstüchtigkeit der SSK auch bei sicherheitstechnisch wichtigen Funktionen anhand deterministischer Sicherheitsanalyse begründen; die Ereignisse schließen betriebliche Transienten, Auslegungsstörfalle, Einwirkung von außen und Naturereignisse ein	29	66	84	21
27	**Auslegungslebensdauer:** Zeitraum, in dem die Funktionsbereitschaft der SSK innerhalb der Akzeptanzkriterien erwartet wird	33	41	91	58
28	**Außerbetriebnahme:** endgültige Abschaltung von SSK	94	90	81	8
29	**Austausch:** Ausbau von nicht abgenutzten, abgenutzten oder ausgefallenen SSK oder eines Teils davon und Einbau entsprechender Teile, die die Funktion innerhalb der ursprünglichen Akzeptanzkriterien erfüllen	92	93	85	15

		Englisch	Französisch	Spanisch	Russisch
30	**Beanspruchung:** Agens oder Mittel, das aus Voreinsatz- oder Einsatzbedingungen entsteht und prompte oder altersbedingte Abnutzung der SSK verursachen kann • *Beispiele*: Wärme, Strahlung, Feuchte, Dampf, Chemikalien, Druck, Vibrationen, seismische Bewegung, elektrische oder mechanische Schwingungen	102	67	1	85
31	**Bedingung:** umgebender Zustand oder Einfluß, der auf die SSK einwirken kann	19	8	12	75
32	**Bedingungen durch Auslegungsereignisse:** Einsatzbedingungen, die durch Auslegungsereignisse herbeigeführt werden	30	11	23	77
33	**Bedingungen vor Inbetriebnahme:** Synonym für *Voreinsatzbedingugen*	81	10	16	52
34	**beschleunigte Alterung:** künstliche Alterung, bei der durch Simulation der natürlichen Alterung in verkürzter Zeit die gleichen Alterungswirkungen erreicht werden, wie unter Betriebsbedingungen über einen längeren Zeitraum	1	101	47	74
35	**bestimmungsgemäßer Betrieb:** Normalbetrieb, einschliesslich gestörtem Betrieb, aber ausschließlich Auslegungsstörfällen oder Erdbeben	70, 71	12	19	95
36	**Betriebsdauer:** Zeitraum ab Inbetriebnahme der SSK bis zum jeweiligen Zeitpunkt	109	47	87	46
37	**Diagnose:** Prüfung und Bewertung der Daten entweder zur Ermittlung des Zustandes der SSK oder der Ursachen für den Zustand	36, 37	35	37	9
38	**Eigenschaft** Merkmal oder Attribut von SSK • *Beispiele*: Form, Abmessungen, Gewicht, Indikatoren für Zustand oder Funktion, realisierte Funktionsfähigkeit oder mechanische, chemische oder elektrische Eigenschaften	14	5	9	91
39	**eigentlicher Grund:** tatsächliche Ursache für den beobachteten Zustand der SSK, deren Beseitigung die Wiederkehr des Zustands ausschließt	96	7	11	25
40	**Einsatzbedingungen:** Zustände oder Einflüsse im Verlaufe der Standzeit von SSK, einschließlich der (normalen und gestörten) Betriebszustände, der Auslegungsereignisse und auslegungsüberschreitender Zustände	98	14	21	60
41	**Einsatzdauer:** Synonym für *Standzeit*	53, 99, 110	38	92	62

		Englisch	Französisch	Spanisch	Russisch
42	**Fehlereffektanalyse:** systematische Ermittlung und Dokumentation der möglichen Ausfallarten und -wirkungen auf die SSK	48	4	7	1
43	**Funktionsbedingungen:** Einflüsse auf SSK bedingt durch Ausführen der Auslegungsfunktionen (Betrieb von System oder Komponente und Belastung einer Konstruktion) • *Beispiele*: – für die Rückschlagklappe: Betriebszyklen und Prellen; – für das Abblaseventile am Reaktordruckbehälter: Reaktorkühlmitteldruck, Durchflußgeschwindigkeit und Temperaturanstieg im Kühlmittel	50, 72	16	24	90
44	**Funktionsindikator:** Zustandsindikator, der direkt die aktuelle Funktionsfähigkeit von SSK innerhalb der Akzeptanzkriterien angibt	51, 74	77	64	89
45	**geplante Standzeit:** Zeitraum, für den anhand von Prüfungen, Analysen oder Erfahrungen nachgewiesen ist, daß die Funktion der SSK innerhalb der Akzeptanzkriterien unter den spezifizierten Betriebsbedingungen gewährleistet ist, und die Sicherheitsfunktionen der SSK im Auslegungsstörfall oder bei Erdbeben erfüllt werden	83	40	76	79
46	**gestörter Betrieb:** Synonym für *anomale Betrieb*	40	17	25	76
47	**installierte Lebensdauer:** Zeitraum zwischen Einbau und Außerbetriebnahme der SSK	55	37	94	68
48	**Instandhaltung:** Kombination direkter und begleitender Maßnahmen zur Erkennung, Vermeidung oder Begrenzung der Abnutzung von SSK oder Wiederherstellung der auslegungskonformen Funktion nach Ausfall der SSK	61	79	67	71
49	**Instandsetzung:** Maßnahmen, die durch Wiederherstellung, Reparatur, Überholung oder Austausch die Funktionsfähigkeit der ausgefallenen SSK innerhalb der Akzeptanzkriterien wieder herstellt	24	80	68	18
50	**kombinierte Wirkungen:** resultierende Änderung der Eigenschaften der SSK, die durch zwei oder mehr Beanspruchungen verursacht wird	15	48	39	65
51	**künstliche Alterung:** Simulation der natürlichen Alterungswirkungen an den SSK durch Aufbringen von Beanspruchungen, die Voreinsatz- und Einsatzbedingungen der Anlage nachbilden, jedoch bei möglicherweise anderer Intensität, Dauer und Einwirkungsart	12	102	48	17

	Englisch	Französisch	Spanisch	Russisch
66 periodische Instandhaltung: Form der vorbeugenden Instandhaltung wie Wartung, Austausch von Teilen, Kontrolle oder Test in festgelegten Intervallen von Kalenderzeit, Betriebsdauer oder Belastungszyklen	75	81	69	45
67 planmäßige Instandhaltung: Form der vorbeugenden Instandhaltung wie Nachrüstung oder Austausch, die geplant ist und vor dem nicht mehr akzeptierbaren Verschleiß der SSK stattfindet	76	84	70	48
68 Prüfung: Beobachtung oder Messung von Zustandsindikatoren unter kontrollierten Bedingungen zum Nachweis, daß die SSK ihren Akzeptanzkriterien genügen	108	54	43	19
69 Prüfung nach Instandhaltungsarbeiten: Überprüfung nach Instandhaltungsarbeiten zur Bestätigung der richtigen Ausführung und Sicherstellung der anschließenden Funktionsweise der SSK innerhalb der Akzeptanzkriterien	77	55	44	20
70 Reparatur: Maßnahmen zur Wiederherstellung eines akzeptierbaren Zustandes der ausgefallenen SSK	91	95	79	61
71 Restauslegungslebensdauer: Zeitraum zwischen einem bestimmten Zeitpunkt und der geplanten Außerbetriebnahme von SSK	87	42	96	37
72 Restlebensdauer: Zeitraum zwischen einem bestimmten Zeitpunkt und der Außerbetriebnahme von SSK	88, 89, 90, 93	44	95	38
73 Revision: Synonym für *wiederkehrende Prüfung*	52	78	66	92
74 Simultanwirkungen: kombinierte Effekte durch gleichzeitig einwirkende Belastungen	101	50	42	36
75 Standzeit: Zeit zwischen Inbetrieb- und Außerbetriebnahme von SSK	53, 99, 110	38	92	62
76 synergistische Wirkungen: Änderung der Eigenschaften von SSK, die durch die Wechselwirkung gleichzeitig wirkender Beanspruchungen hervorgerufen wird; zu unterscheiden von Änderungen, die durch Überlagerung unabhängig wirkender Beanspruchungen entstehen	107	51	41	63
77 systematischer Fehler: Synonym für *Ausfall aus gemeinsamer Ursache* und *Ausfall gleicher Art*	16, 17	–	–	–

	Englisch	Französisch	Spanisch	Russisch
78	**Test:** Synonym für *Prüfung*	108	54	43	19
79	**Totalausfall:** vollständiger Funktionsverlust	13, 18	25	56	51
80	**Trendbildung der Ausfälle:** Registrierung, Analyse und Extrapolation von Betriebsausfällen in SSK als Funktion unabhängiger Parameter (in der Regel Zeit oder Belastungszyklen)	49	58	86	55
81	**Überholung:** Instandsetzungsarbeiten größeren Umfangs, Nachrüstung oder beides	73, 85	96	83	22
82	**Überwachung (1):** Beobachtung oder Messung der Zustands- oder Funktionsindikatoren, um die Einhaltung der Akzeptanzkriterien der SSK nachzuweisen	103	98	65	94
83	**Überwachung(2):** kontinuierliche Überwachung der Anlagenbedingungen während Betrieb oder Stillstand	104	–	99	23
84	**Überwachungsanforderung:** Prüfung, Kalibrierung oder Inspektion zur Sicherstellung der erforderlichen Qualität der SSK, die den sicheren Betrieb innerhalb vorgegebener Grenzen gewährleistet	105	89	80	31
85	**Umgebungsbedingungen:** Zustände der Umgebung von SSK • *Beispiele*: Temperatur, Strahlung, Feuchte im Containment, während Normalbetrieb oder bei Störfällen	38	9	15	78
86	**Versagen:** Synonym für *Ausfall*	42, 62	23	54	39
87	**Versagensart:** Art und Weise oder Zustand, in dem SSK ausfallen • *Beispiele*: Offenbleiben (Ventil), Kurzschluß (Kabel), Lagerfraß (Motor), Leckage (Ventil, Behälter oder Containment), Durchflußbehinderung (Rohr oder Ventil), Ausfall der Anregung für die Abschaltstäbe (Reaktorschutzsystem), Riß oder Bruch (Struktur)	47	87	75	4
88	**Verschleiß:** prompte oder allmähliche Verschlechterung der Eigenschaften der SSK, die deren Funktionsfähigkeit innerhalb der Akzeptanzkriterien beeinträchtigen könnte	25, 35	29	30	80
89	**Voralterung:** Simulation der natürlichen Alterung der SSK durch kombinierte künstliche und natürliche Alterung	4, 78	60	52	29

		Englisch	Französisch	Spanisch	Russisch
90	**vorauseilende Instandhaltung:** Form der vorbeugenden Instandhaltung, die kontinuierlich oder in Intervallen durchgeführt wird in Abhängigkeit von beobachtetem Zustand, Diagnose oder Trendbildung von Zustandsindikatoren; die Ergebnisse betreffen gegenwärtige und künftige Funktionsweisen oder Art und Zeitplan der planmäßigen Instandhaltung	79	83	71	59
91	**vorbeugende Instandhaltung:** Maßnahmen zur Erkennung, zum Ausschluß oder zur Begrenzung der Abnutzung an funktionstüchtigen SSK oder zur Lebensdauerverlängerung durch Beschränkung der Abnutzung und Ausfälle auf ein akzeptables Maß; man unterscheidet drei Arten der vorbeugenden Instandhaltung: wiederkehrend, vorauseilend und planmäßig	82	82	72	47
92	**Voreinsatzbedingungen:** Zustände der SSK oder Einflüsse darauf vor Inbetriebnahme • *Beispiele*: Herstellung, Lagerung, Transport, Montage und Prüfungen vor Inbetriebnahme	81	10	16	52
93	**vorzeitige Alterung:** Alterung im Betrieb, die schneller als erwartet fortschreitet	80	105	51	53
94	**Wartung:** Routinearbeiten (einschließlich reinigen, einstellen, kalibrieren und wechseln der Verbrauchsstoffe) zur Erhaltung oder Verlängerung der Nutzungsdauer von SSK	100	52	82	70
95	**Wartung und Instandsetzung** Synonym für **Instandhaltung**	61	–	–	–
96	**Wiederholungsprüfung:** Prüfung zur Ermittlung der Betriebsbereitschaft von Komponenten und Systemen	54	53	45	93
97	**wiederkehrende Prüfung:** Methoden und Vorgehensweisen, die Integrität der sicherheitstechnisch bedeutsamen Strukturen und druckführenden Komponenten des Kernkraftwerks zu gewährleisten	52	78	66	92
98	**Zufallsausfall:** jeder Ausfall, über dessen Ursache oder Mechanismus keine Eintrittsvoraussage gemacht werden kann	84	24	55	64
99	**Zustand:** Zustand oder Beschaffenheit der SSK, der die Fähigkeit zur Erfüllung einer Auslegungsfunktion beeinträchtigen kann	20	–	13	66

		Englisch	Französisch	Spanisch	Russisch
100	**Zustandsindikator:** Eigenschaft, die beobachtet oder gemessen werden kann oder sich zur Trendbildung eignet, zur Feststellung der gegenwärtigen und künftigen Funktionsfähigkeit der SSK innerhalb der Akzeptanzkriterien	**21**	**75**	**62**	**50**
101	**Zustandsüberwachung:** Beobachtung, Messung oder Trendbildung von Zustands- oder Funktionsindikatoren bezüglich unabhängiger Parameter (in der Regel Zeit oder Belastungszyklus) zur Anzeige der gegenwärtigen und künftigen Funktionsfähigkeit der SSK innerhalb der Akzeptanzkriterien	**22, 23**	**97**	**28**	**24**

BASISIDEEN ZUM ALTERUNGSBEGRIFF

Ursachen für Verschleiß

- **Einsatzbedingungen** sind alle auftretenden Bedingungen, die Einfluß nehmen auf SSK. Sie umfassen den bestimmungsgemäßen Betrieb (**Normalbetrieb, gestörter Betrieb** und betriebliche Transienten) und Auslegungsstörfälle.

- **Auslegungsbedingungen** sind hypothetische Zustände die i.a. Sicherheitszuschläge über die erwarteten Einsatzbedingungen hinaus enthalten.

Lebensdauer

- Die **Standzeit** ist der tatsächliche Zeitabschnitt, in dem die SSK ihre Funktion erfüllen, sie kann sich unterscheiden von der erwarteten Standzeit, der **Auslegungslebensdauer**.

- Das **Alter** von SSK (gerechnet ab Fertigung) kann sich von der **Betriebsdauer** unterscheiden (gerechnet ab Inbetriebnahme der SSK).

Verschleiß/Alterung

- Der **Verschleiß** erfolgt allmählich (Alterung) oder prompt (keine Alterung).

- **Alterungsbedingter Verschleiß** entsteht bei **bestimmungsgemäßem Betrieb**, sowohl durch **Umgebungsbedingungen** wie Temperatur und Strahlung als auch **Funktionsbedingungen**, wie z.B. relative Bewegung zwischen Teilen. Bestimmungsgemäßer Betrieb erzeugt **normale** oder **anomale Beanspruchungen**.

- **Auslegungsereignisse** schließen betriebliche Transienten ein (die zur **Alterung** beitragen können) sowie Auslegungsstörfälle und Erdbeben (die prompten, nicht-alterungsbedingten **Verschleiß** verursachen).

- **Abgenutzte Zustände** - aber noch keine **Ausfälle** - sind bis zu einem gewissen Grad akzeptabel. Im Fall eines **Ausfalls durch Verschleiß** werden zwar die **Akzeptanzkriterien** verletzt, aber eine Teilfunktion bleibt erhalten. Bei **Totalausfall** gibt es keine Funktion mehr.

Ausfall

- **Ausfälle** entstehen durch Ursachenketten, nicht durch eine Ursache allein. Bei **Ausfall durch Alterung** ist die aktuelle Ausfallursache ein **Alterungsmechanismus**. Der **eigentliche Grund** kann außerhalb der Alterung liegen.

- **Vorzeitige Alterung** kann zu Betriebs**ausfällen** der SSK führen. Der Begriff **beschleunigte Alterung** sollte für die **künstliche Alterung** reserviert bleiben, die normalerweise im Labor erzeugt wird.

- **Ausfallanalysen** identifizieren **Ausfallursachen**, **Ausfallmechanismen** und die **Versagensart**. Jeder dieser Begriffe hat einen anderen Inhalt. Der **eigentliche Grund** von **anomalem alterungsbedingten Verschleiß** und von Ausfällen ist nicht die **Alterung** sondern eher menschliches Versagen.

Instandhaltung/Zustandbewertung

- **Instandhaltung** ist ein umfassender Begriff und beinhaltet die **Instandsetzung** und **Wartung** (**vorbeugende Instandhaltung**). **Instandhaltung** kann von Wartungs–, technischem oder Betriebspersonal durchgeführt werden. **Vorbeugende Instandhaltung** umfaßt **vorauseilende Instandhaltung** wie **Tests** und **Zustandsüberwachung**.

- **Reparaturen** werden nur an ausgefallenen SSK durchgeführt, **Nachrüstung** nur an nicht ausgefallenen SSK. Eine **Überholung** ist eine umfangreiche Reparatur und/oder Nachrüstung.

Abb 1. **Beziehungen zwischen den Begriffen**

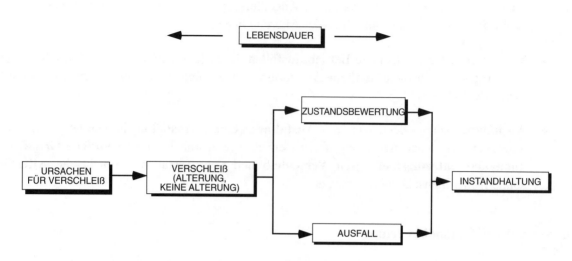

Hinweis: Die Zustandsbewertung stellt die Entscheidungsgrundlage zur Planung der Instandhaltung dar.

Abb 2 Beziehungen zwischen den Einsatzbedingungen

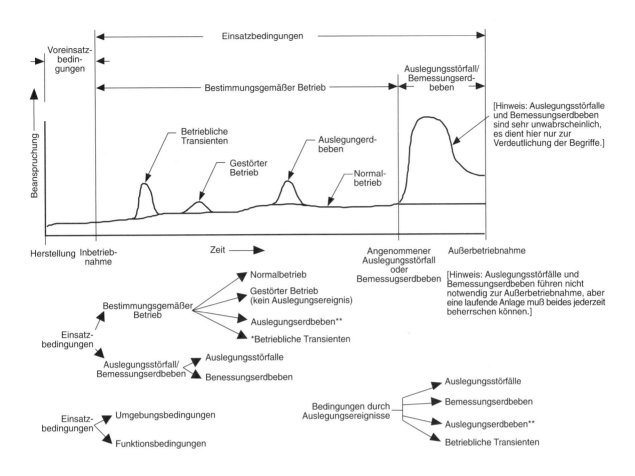

* Betriebliche Transienten können auch durch Störungen ausgelöst sein, die Anlage ist dafür ausgelegt. Die Auslegungsbedingungen umfassen alle Einsatzbedingungen jenseits des gestörten Betriebs.
** In Deutschland ist nur noch das Bemessungserdbeben gefordert.

Abb. 3. Beziehungen zwischen Alterungsbegriffen

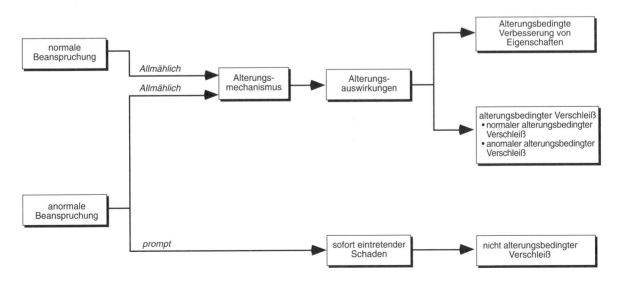

Abb. 4. Beziehungen zwischen Eckpunkten während der Lebendauer, aufgetragen auf der Zeitachse

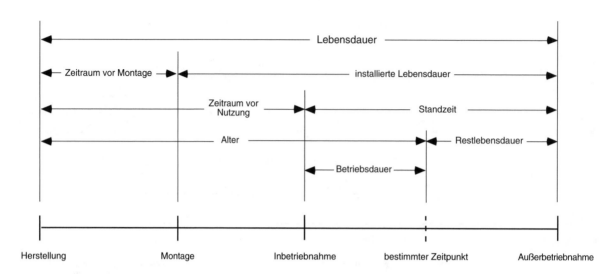

Hinweis: Diese Begriffe stellen Eckpunkte der vergangenen oder künftigen Entwicklung eines systems, einer Struktur oder einer Komponente dar.

Abb. 5 Beziehungen zwischen Auslegungeseckpunkten während der Auslegungslebensdaur

A. Ursprüngliche Annahmen zur Auslegungslebensdauer

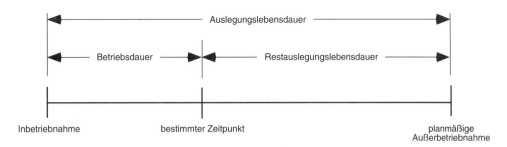

B. Mögliche Verlängerung der Auslegungslebensdauer auf Grund revidierter Annahmen (je nach Einschätzung)

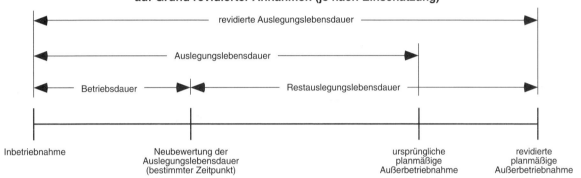

C. Mögliche Verkürzung der Auslegungslebensdauer auf Grund revidierter Annahmen (je nach Einschätzung)

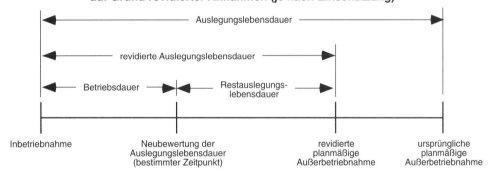

Abb. 6 Beziehung zwischen Begriffen zum Stichwort Ausfall

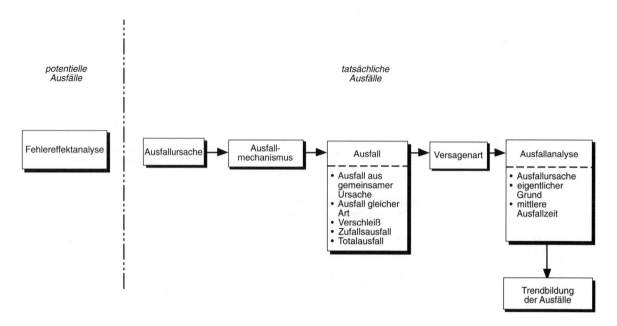

Abb. 7. Beispiele für Ursachen, Mechanismen und Erscheinungsformen

Definitionen zu ausfallursachen

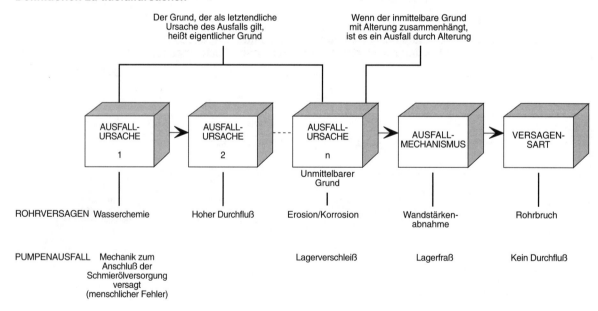

Abb 8.   **Begriffe zur Instandhaltung**

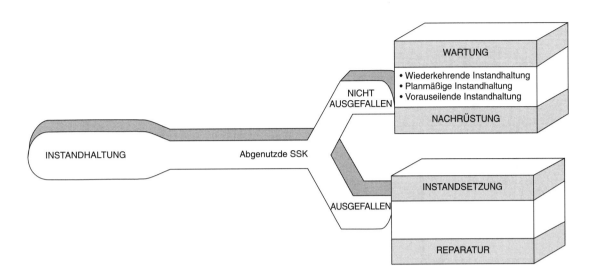

Abb. 9 Beziehungen zwischen Instandhaltungsbegriffen

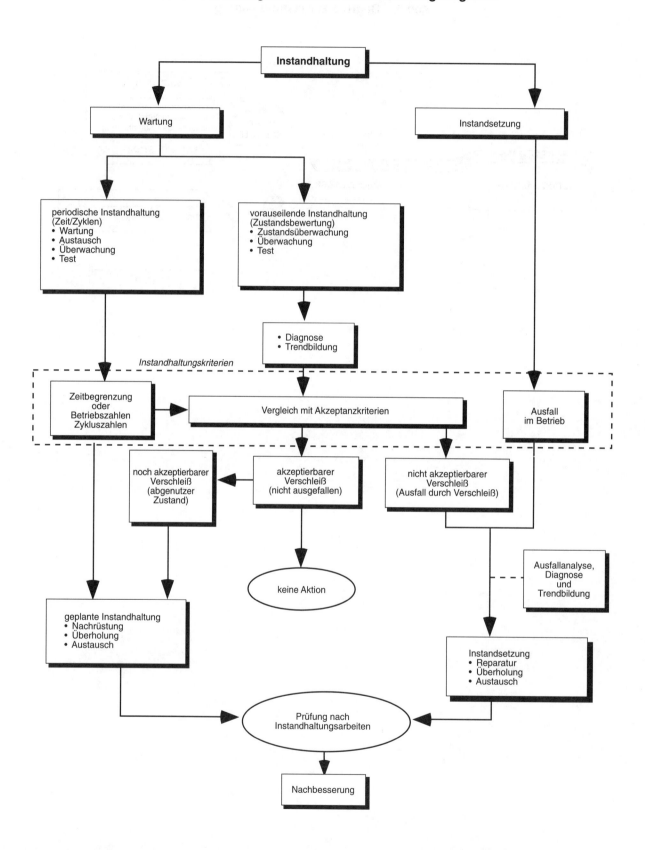

GLOSARIO

ENVEJECIMIENTO DE PLANTAS NUCLEARES

Un glosario útil para el entendimiento y la gestión

del envejecimiento de systemas, estructuras

y componentes de plantas nucleares

POR QUE UNA TERMINOLOGIA COMUN DE ENVEJECIMIENTO?

A medida que la vida de servicio de las plantas nucleares aumenta, se da más atención a los posibles mal entendidos relacionados con la degradación por envejecimiento de sistemas, estructuras o componentes (SSCs). La terminología común en envejecimiento ha sido desarrollada para mejorar el entendimiento del fenómeno de envejecimiento, para facilitar el reporte de datos de fallos relevante y para promover interpretaciones uniformes de standards y regulaciones relativas a envejecimiento.

La terminología debe ser útil en temas relacionadas con gestion de vida y de envejecimiento. La gestión de vida puede minimizar costos de operación y mantenimiento y puede soportar la opción de extension de vida de las plantas de 40 a 60 años. Aun mas importante es el hecho de que una gestíon efectiva de envejecimiento contribuye al mantenimiento de los margenes adequados de seguridad.

Reconociendo la importancia de una comunicación clara en estas areas, representantes de diversos sectores de la indústria han desarroyado un vocabulario uniforme de terminos relacionados a envejecimiento.

En vista de los beneficios que pueden ser obtenidos, se recomienda el uso de la terminología común de envejecimiento en documentos técnicos, informes de investigación, regulaciones futuras, y otra documentación relacionada con el envejecimiento de plantas nucleares. La documentación debería especificar el uso de la terminología común de envejecimiento excepto donde sea señalado. Excepciones apropiadas serían aquellos casos en los cuales el escritor opta por usar definiciones de standards y regulaciones existentes.

Los beneficios màs importantes del uso de la terminología común en envejecimiento son:

- Mejorar el reporte y la interpretación de datos de planta en sistemas, estructuras y componentes;

- Mejorar la interpretación y el cumplimiento con códigos, standards y regulaciones relacionadas con el envejecimiento de plants nucleares.

Por que este glosario?

La NEA (Agencía para la Energía Nuclear de la OCDE) has publicado este glosario en cooperación con la CEC (Comisión de las Comunidades Europeas) y la IAEA (Organismo Internacional de Energía Atómica) en conjunto, como una rápida referencia para facilitar y motivar ampliamente el uso de la terminología común de envejecimiento. El objetivo es proveer al personal de planta (y otros interesados en el envejecimiento) con un conjunto común de teminos que tienen uniformidad, significado amplio en la industria y para facilitar la discusión entre expertos de diferentes paises.

En cada sección de idiomas, los terminos son listados en orden alfabético con un número secuencial. Estos numeros son repetidos en la seccion del idioma Ingles, permitiendo una referencia cruzada entre todos los idiomas. El guión en las columnas de números de referencia indica que no hay un término que corresponde en el idioma respectivo. Note que los guiones aparecen solo en las columnas de términos sinónimos.

En cada idioma, el Glosario comienza con un vision general de todos los terminos agrupados en seis categorias. Esto es seguido por un listado de terminos en orden alfabético, definiciones y algunos cuantos ejemplos. Las últimas páginas de cada sección contienen diagramas y una lista de ideas claves para ayudar a entender la terminología.

Agradecimiento

Este Glosario es una representación muy cercana a la publicacion (BR-101747) de EPRI (Instituto de Investigación para la Energía Eléctrica) a quien nosotros agadecemos por el soporte y la ayuda por hacer disponible esta versión internacional. El Glosario de EPRI fué desarrollado con contribuciones de algunas empresas electricas de servicio público de Los Estados Unidos, la Comisión de Regulación Nuclear de Los Estados Unidos, el Instituto de Energía Nuclear y Los Laboratorios Nacionales de Los Estados Unidos.

Terminología Básica de Envejecimiento Ordenada por Categorias

DEGRADACIÓN	
CAUSAS DE LA DEGRADACIÓN	**DEGRADACIÓN/ENVEJECIMIENTO**
condición • condiciones de servicio • condiciones antes del servicio • condiciones ambientales • condiciones funcionales • condiciones de operación • condiciones normales • condiciones inducidas por error • suceso base de diseño • condiciones de suceso base de diseño • condiciones de diseño agente de envejecimiento • agente de envejecimiento normal • agente de envejecimiento inducido por error • agente de envejecimiento originado por un suceso base de diseño	característica condición • condición degradada envejecimiento • envejecimiento natural • envejecimiento prematuro • envejecimiento normal • envejecimiento artificial • envejecimiento acelerado • envejecimiento simulado mecanismo de envejecimiento efectos del envejecimiento • efectos combinados • efectos simultáneos • efectos de sinergía degradación • degradación por envejecimiento • degradación por envejecimiento normal • degradación por envejecimiento inducida por error valoración del envejecimiento
CICLO VITAL	
VIDA	**FALLO**
edad tiempo en servicio vida • vida instalada • vida de servicio • vida remanente • vida de diseño • vida remanente de diseño • período cualificado retirada	fallo • fallo parcial • fallo completo • fallo aleatorio • fallos de causa común • fallos de modo común • desgaste causa del fallo • causa raíz • mecanismo del fallo • modo del fallo análisis de fallos • evaluación de fallos • modos de fallo y análisis de efectos • tendecias del fallo tiempo medio entre fallos
GESTIÓN DEL ENVEJECIMIENTO	
MANTENIMIENTO	**EVALUACIÓN DE LA CONDICIÓN**
gestión del envejecimiento gestión de vida útil mantenimiento • mantenimiento preventivo • mantenimiento periódico • mantenimiento planificado • mantenimiento correctivo reparación reconstrucción revisión general sustitución revisión ensayo de postmantenimiento reajuste	mantenimiento predictivo inspección en servicio ensayo en servicio inspección vigilancia requisitos de vigilancia control de la condición indicador de condición indicador funcional ensayo diagnosis criterio de aceptación

		inglés	francés	alemán	ruso
1	**agente de envejecimiento:** agente o estímulo derivado de condiciones antes del servicio y condiciones de servicio y que provoca la degradación inmediata o el envejecimiento de un SEC[1.] • *Ejemplos*: calor, radiación, humedad, vapor, condiciones ambientales químicas, presión, vibración, terremoto, transitorios eléctricos y desgaste mecánico	102	67	30	85
2	**agente de envejecimiento inducido por error:** agente de envejecimiento originado por condiciones inducidas por error que, inmediatamente o a plazo, provoca una degradación superior a la que provocarían los agentes de envejecimiento normales	41	69	12	86
3	**agente de envejecimiento normal**: agente de envejecimiento originado en condiciones normales que puede provocar en un SEC mecanismos y efectos de envejecimiento	69	70	64	88
4	**agente de envejecimiento originado por un suceso base de diseño:** agente de envejecimiento originado por sucesos base de diseño que, inmediatamente o a plazo, causa una degradación superior a la que provocarían los agentes de envejecimiento normales	31	68	24	87
5	**análisis de causa raíz:** sinónimo de *análisis de fallos*	97	2	20	2
6	**análisis de fallos:** proceso en que sistemáticamente se determinan y documentan el modo, el mecanismo, las causas y la causa raíz del fallo de un SEC	43	3	20	3
7	**análisis de modos de fallo y sus efectos:** proceso en que sistemáticamente se determinan y documentan los modos de fallo previsibles y sus efectos sobre los SEC	48	4	42	1
8	**avería:** sinónimo de *fallo completo*	13	–	79	7
9	**característica:** propiedad o atributo de un SEC (forma, dimensión, peso, indicador de condición, indicador de función, rendimiento o cualquier atributo mecánico, químico o eléctrico)	14	5	38	91
10	**causa del fallo:** circunstancia originada durante el diseño, la fabricación, el ensayo o el uso, que acaban provocando un fallo	44	6	23	54

---

1. SEC = sistema, estructura o componente

		inglés	francés	alemán	ruso
11	**causa raíz:** razón principal de una condición observada en un SEC cuya corrección evitaria la repetición de dicho estado análisis	96	7	39	25
12	**condición (1):** ambiente fisico o de otra clase que pueden afectar a un SEC	19	8	31	75
13	**condición (2):** estado o nivel de las características de un SEC que puede condicionar su capacidad de cumplir una función de diseño	20	–	99	66
14	**condición degradada:** situación de un SEC en el que sin haberse producido el fallo sea aconsejable una operación de mantenimiento planificado	26	61	1	84
15	**condiciones ambientales:** estados físicos que configuran el medio ambiente de un SEC • *Ejemplos*: temperatura, radiación, y humedad en la contención durante la operación normal o accidentes	38	9	85	78
16	**condiciones antes del servicio:** estados físicos o ambientales que afectan a un SEC antes de su puesta en funcionamiento (p.e.: fabricación, almacenamiento, transporte, instalación y ensayo preoperacional)	81	10	33, 92	52
17	**condiciones base de diseño:** sinónimo de *condiciones de diseño*	28	–	25	–
18	**condiciones de diseño:** condiciones de diseño que se especifican para un SEC conforme con las normas y directrices existentes y que se incluyen habitualmente en las especificaciones técnicas (que prevén generalmente un margen de maniobra que rebasa las condiciones de servicio previsibles)	32	18	25	57
19	**condiciones de operación:** condiciones de servicio, incluídas las condiciones normales y las condiciones inducidas por error, anteriores al inicio de un accidente base de diseño o a un terremoto	70	12	35	95
20	**condiciones de operación normal:** sinónimo de *condiciones de operación*	68	20	61	35
21	**condiciones de servicio:** estados físicos o ambientales presentes durante la vida de servicio de un SEC, incluídas las condiciones de operación (normales e inducidas por error), las condiciones de sucesos base de diseño y las condiciones de sucesos posterior al diseño	98	14	40, 52	60
22	**condiciones de servicio de diseño:** sinónimo de *condiciones de diseño*	34	13	25	–

		inglés	francés	alemán	ruso
23	**condiciones de suceso base de diseño:** condiciones de servicio producidas por sucesos base de diseño	**30**	**11**	**32**	**77**
24	**condiciones funcionales:** conjunto de las influencias sufridas por un SEC que se derivan del desarrollo de sus funciones de diseño (funcionamiento de un sistema o componente y carga de una estructura) • *Ejemplos*: – para una válvula de retención - oscilación y vibración operacional – para una válvula de seguridad de la vasija de reactor - presión del refrigerante del reactor, velocidades altas de caudal, e incremento de temperaturas en el refrigerante del reactor	**50**	**16**	**43**	**90**
25	**condiciones inducidas por error:** condiciones adversas de preservicio o de servicio debidas a errores de diseño, fabricación, funcionamiento o mantenimiento	**40**	**17**	**14, 46**	**76**
26	**condiciones normales:** condiciones de funcionamiento de un SEC correctamente diseñado, fabricado, instalado, operado y mantenido, con exclusión de las condiciones de suceso base de diseño	**67**	**19**	**61**	**34**
27	**condiciones operacionales:** sinónimo de *condiciones funcionales*	**72**	**21**	**43**	**–**
28	**control de la condición:** observación, medida o tendencia de los indicadores de la condición o del funcionamiento consideradas en relación a un parámetro independiente (tiempo o ciclos, generalmente) que indica la capacidad presente o futura que tiene un SEC de funcionar con arreglo a los criterios de aceptación	**22**	**97**	**101**	**24**
29	**criterio de aceptación:** límite especificado de un indicador funcional o de condición con que se evalúa la capacidad que tiene un SEC para cumplir con sus funciones de diseño	**2**	**22**	**3**	**26**
30	**degradación:** deterioro inmediato o gradual de las características de un SEC que merma su capacidad de funcionar con arreglo a los criterios de aceptación	**25**	**29**	**2, 88**	**80**
31	**degradación por envejecimiento:** conjunto de efectos de envejecimiento que pueden impedir que el SEC cumpla los criterios de aceptación • *Ejemplos*: reducción en el diámetro de un eje por desgaste, pérdida de resistencia de material por fatiga o envejecimiento térmico, hinchamiento de compuestos encapsulados, y pérdida de resistencia dieléctrica o agrietamiento de aislamiento	**7**	**31**	**7, 8**	**81**

	inglés	francés	alemán	ruso
32	**degradación por envejecimiento inducida por error**: degradación por envejecimiento producida por condiciones inducidas por error	39	32	13	83
33	**degradación por envejecimiento normal:** degradación por envejecimiento producida en condiciones normales	66	33	63	82
34	**degradación relacionada con el envejecimiento:** sinónimo de *degradación por envejecimiento*	11	30	7, 8	–
35	**desgaste:** fallo provocado por un mecanismo de envejecimiento	111	99	17	16
36	**deterioro:** sinónimo de *degradación*	35	34	2, 88	–
37	**diagnosis:** examen y evaluación de datos con que se determina la condición de un SEC o las causas de la misma	36	35	37	9
38	**edad:** tiempo transcurrido entre la fabricación de un SEC y un momento dado	3	1	4	5
39	**efectos combinados:** alteraciones netas de las características de un SEC debidas a la actuación de dos o más agentes de envejecimiento	15	48	50	65
40	**efectos de envejecimiento:** cambios netos que producen el tiempo y el uso en las características de un SEC debidos a los mecanismos del envejecimiento • *Ejemplos*: efectos negativos - ver *degradación por envejecimiento*; efectos positivos - aumento de la resistencia del hormigón por curado; disminución de la vibración por asentamiento de la maquinaria rotativa	8	49	6	96
41	**efectos de sinergía:** conjunto de alteraciones de un SEC producidas exclusivamente por la acción simultánea de varios agentes de envejecimiento, que deben distinguirse de las alteraciones producidas por la superposición de la acción aislada de cada agente de envejecimiento	107	51	76	63
42	**efectos simultáneos:** efectos combinados de varios agentes de envejecimiento que actúan simultáneamente	101	50	74	36
43	**ensayo:** observación o medida de los indicadores de condición efectuadas en condiciones controladas con que se comprueba que un SEC concreto cumple los criterios de aceptación	108	54	68, 78	19
44	**ensayo de postmantenimiento:** ensayo realizado después de una operación de mantenimiento para comprobar que la misma se ha desarrollado correctamente y que el SEC puede funcionar con arreglo a los criterios de aceptación	77	55	69	20

		inglés	francés	alemán	ruso
45	**ensayo en servicio:** ensayo con que se determina la capacidad operativa de un sistema o componente	**54**	**53**	**96**	**93**
46	**envejecimiento:** proceso general por el que las características de un SEC van cambiando con el tiempo o con el uso	**5**	**106**	**5**	**69**
47	**envejecimiento acelerado:** técnica de envejecimiento artificial con la que se simulan, en un tiempo breve, los efectos del envejecimiento natural producido en un período más largo de servicio (ver también *envejecimiento prematuro*)	**1**	**101**	**34**	**74**
48	**envejecimiento artificial:** simulación de los efectos del envejecimiento natural que se realiza sometiendo los SEC a condiciones que sustituyen a las de preservicio y de servicio pero que pueden ser diferentes en intensidad, duración y forma de aplicación	**12**	**102**	**51**	**17**
49	**envejecimiento natural:** envejecimiento de un SEC producido en condiciones de preservicio y de servicio, incluidas las condiciones inducidas por error	**64**	**103**	**60**	**12**
50	**envejecimiento normal:** envejecimiento natural debido a condiciones no erróneas de preservicio o de servicio	**65**	**104**	**62**	**33**
51	**envejecimiento prematuro:** envejecimiento en servicio que avanza a un ritmo más acelerado de lo previsto	**80**	**105**	**93**	**53**
52	**envejecimiento simulado:** reproducción de los efectos producidos por el envejecimiento natural en un SEC, sometiéndolo a una técnica combinada de envejecimiento artificial y natural	**4**	**60**	**89**	**29**
53	**evaluación de fallos:** conclusión sacada del análisis de fallos	**45**	**63**	**21**	**43**
54	**fallo:** incapacidad o interrupción de la capacidad de un SEC para funcionar con arreglo a los criterios de aceptación	**42**	**23**	**15, 86**	**39**
55	**fallo aleatorio:** cualquier fallo de cuya causa y/o de cuyo mecanismo no se pueda deducir el momento en que se producirá	**84**	**24**	**98**	**64**
56	**fallo completo:** fallo que provoca una pérdida total de funcionamiento del SEC	**18**	**25**	**75**	**51**
57	**fallos de causa común:** dos o más fallos debidos a la misma causa	**16**	**27**	**16, 77**	**41**
58	**fallos de modo común:** dos o más fallos de igual naturaleza debidos a la misma causa	**17**	**26**	**19, 77**	**14**

		inglés	francés	alemán	ruso
59	**fallo parcial:** fallo en el que un indicador funcional no cumple con un criterio de aceptación pero no se pierde del todo la función de diseño	27	71	18	40
60	**gestión de vida útil:** integración de la gestión del envejecimiento y la planificación económica para: (1) optimizar el funcionamiento, el mantenimiento y la vida útil de los SEC; (2) mantener un nivel aceptable de eficacia y seguridad; y (3) obtener el rendimiento máximo de la inversión durante la vida útil de la central	59	72	55	72
61	**gestión del envejecimiento:** conjunto de actuaciones de ingeniería, operación y mantenimiento con las cuales se pretende que la degradación y el desgaste de los SEC no sobrepasen unos límites aceptables • *Ejemplos* de actuaciones de ingeniería: diseño, cualificación y análisis de fallos • *Ejemplos* de actuaciones de operación: inspección, realización de procedimientos de operación dentro de unos límites especificados, y realización de medidas medioambientales	9	74	10	73
62	**indicador de condición:** característica observable, medible o extrapolable que indica - o de la que se deduce - la capacidad presente o futura que tiene un SEC de funcionar dentro de los criterios de aceptación	21	75	100	50
63	**indicador de rendimiento:** sinónimo de *indicador funcional*	74	76	44	49
64	**indicador funcional:** indicador de condición que mide directamente la capacidad presente que tiene un SEC para funcionar con arreglo a los criterios de aceptación	51	77	44	89
65	**inspección:** observación o medida de los indicadores funcionales que se efectúan para comprobar que un SEC concreto funciona con arreglo a los criterios de aceptación	103	98	82	94
66	**inspección en servicio:** conjunto de métodos y operaciones que garantizan la integridad estructural y la resistencia a la presión de componentes relacionados con la seguridad de centrales nucleares	52	78	73, 97	92
67	**mantenimiento:** conjunto de actividades directas o de apoyo con que se detecta, impide o mitiga la degradación de un SEC en funcionamiento, o con que se recomponen a un nivel aceptable las funciones de diseño de un SEC averiado	61	79	48, 95	71

	inglés	francés	alemán	ruso

68 **mantenimiento correctivo:** conjunto de operaciones - reparación, revisión general o sustitución - que restablecen la capacidad de un SEC averiado para funcionar con arreglo a los criterios de aceptación — inglés **24**, francés **80**, alemán **49**, ruso **18**

69 **mantenimiento periódico:** tipo de mantenimiento preventivo en el que, con fechas predeterminadas, tiempo de funcionamiento o número de ciclos, se hace una revisión, se sustituyen piezas y se invigila o ensaya el funcionamiento del sistema — inglés **75**, francés **81**, alemán **66**, ruso **45**

70 **mantenimiento planificado:** tipo de mantenimiento preventivo en que se reconstruyen o sustituyen elementos y que es programado y realizado antes de que pueda producirse la degradación inaceptable de un SEC — inglés **76**, francés **84**, alemán **67**, ruso **48**

71 **mantenimiento predictivo:** tipo de mantenimiento preventivo efectuado con carácter continuo o a intervalos, con el que se controlan, se diagnostican o se analizan los indicadores funcionales o de condición de un SEC; los resultados indican la capacidad funcional presente y futura o el tipo y el calendario del mantenimiento planificado — inglés **79**, francés **83**, alemán **90**, ruso **59**

72 **mantenimiento preventivo:** conjunto de operaciones con que se detecta, evita o mitiga la degradación de un SEC en servicio y que, al mantener la degradación y los fallos dentro de unos límites razonables, se refuerza o prolonga la vida útil del mismo. Hay tres tipos de mantenimiento preventivo: periódico, predictivo y planificado — inglés **82**, francés **82**, alemán **91**, ruso **47**

73 **mecanismo de envejecimiento:** proceso específico por el que el tiempo o el uso alteran gradualmente las características de un SEC — inglés **10**, francés **86**, alemán **11**, ruso **28**

- *Ejemplos*: curado, desgaste, fatiga, deformación permanente por fatiga, erosión, incrustación microbiológica, corrosión, fragilización y reacciones químicas o biológicas

74 **mecanismo del fallo:** proceso físico que culmina en el fallo — inglés **46**, francés **85**, alemán **22**, ruso **27**

- *Ejemplos*: agrietamiento de un cable aislante frágil (relacionado a envejecimiento); un objeto obstruyendo flujo (no relacionado a envejecimiento)

75 **modo del fallo** manera o situación en que falla un SEC — inglés **47**, francés **87**, alemán **87**, ruso **4**

- *Ejemplos*: atascamiento abierto (válvula), corto de tierra (cable), detención de cojinete (motor), fúga (válvula, recipiente, contención), paro de flujo (tubería o válvula), fallo de produción de señal de bajada de varras de control (sistema de protección del reactor), grieta o ruptura (estrutura)

		inglés	francés	alemán	ruso
76	**período cualificado:** período durante el que, según demuestran los ensayos, los análisis o la experiencia, un SEC es capaz de funcionar en las condiciones especificadas de operación, manteniendo su capacidad de cumplir con las funciones de seguridad en caso de accidente base de diseño o en caso de terremoto, dentro del rango de los criterios de aceptación	83	40	45	79
77	**reajuste:** operación que corrige una fabricación, una instalación o un mantenimiento defectuosos	95	91	58	11
78	**reconstrucción:** acciones planificadas para mejorar las condiciones de un SEC antes de que falle	86	94	59	6
79	**reparación:** conjunto de operaciones con las que se restituye un SEC a una condición aceptable	91	95	70	61
80	**requisitos de vigilancia:** ensayo, calibración o inspección con que se mantiene el nivel necesario de calidad de los sistemas y componentes, y se garantiza que las instalaciones funcionen dentro de los límites de seguridad sin sobrepasar las condiciones limitativas de operación (Utilizar este término sólo cuando se den connotaciones normativas y legales específicas)	105	89	84	31
81	**retirada:** desactivación definitiva de un SEC	94	90	28	8
82	**revisión:** conjunto de operaciones de rutina (limpieza, ajuste, calibrado y sustitución de elementos consumibles) que mantienen o prolongan la vida útil de un SEC	100	52	94	70
83	**revisión general:** reparación generalizada, reconstrucción o ambos	73	96	81	22
84	**suceso base de diseño:** un suceso que se especifica para establecer el cumplimiento aceptable en condiciones de operación normal y, mediante análisis de seguridad deterministas, de las funciones de seguridad de los SEC. Se consideran como tales los transitorios anticipados, los accidentes base de diseño, los sucesos externos y los fenómenos naturales	29	66	26	21
85	**sustitución:** reemplazo de un SEC intacto, degradado o averiado, o de una parte del mismo, que se efectúa para instalar en su lugar otro SEC u otros elementos que puedan funcionar dentro del rango de aceptación originales	92	93	29	15
86	**tendencias del fallo:** registro, análisis y previsión de fallos en servicio de un SEC con relación a parámetros independientes (tiempo o ciclos generalmente)	49	58	80	55
87	**tiempo en servicio:** tiempo transcurrido desde la puesta en funcionamiento del SEC hasta un momento dado	109	47	36	46

	inglés	francés	alemán	ruso
88 tiempo medio entre fallos: media aritmética de los intervalos transcurridos entre los fallos de un SEC	63	88	57	67
89 valoración del envejecimiento: evaluación de la información necesaria para determinar los efectos del envejecimiento sobre la capacidad presente y futura de funcionamiento de los SEC para cumplir los criterios de aceptación	6	65	9	44
90 vida: período transcurrido entre la fabricación y la retirada de un SEC	56	36	53	13
91 vida de diseño: período de tiempo durante el que se calcula que un SEC debe funcionar con arreglo a los criterios de aceptación	33	41	27	58
92 vida de servicio: período que va desde la puesta en funcionamiento de un SEC hasta su retirada	99	38	41, 65,75	62
93 vida en servicio remanente: sinónimo de *vida remanente*	89	39	72	–
94 vida instalada: período transcurrido entre el momento de la instalación y el momento de retirada de un SEC	55	37	47	68
95 vida remanente: período que abarca desde un momento dado hasta el momento concreto de retirada de un SEC	88	44	72	38
96 vida remanente de diseño: período que abarca desde un momento dado hasta el momento previsto de retirada de un SEC	87	42	71	37
97 vida útil: sinónimo de *vida de servicio*	110	45	41, 65, 75	–
98 vida útil remanente: sinónimo de *vida remanente*	90	46	72	–
99 vigilancia: control continuo de las condiciones de la planta durante su operación o en parada	104	–	83	23
100 vigilancia de la condición: sinónimo de *control de la condición*	22	–	–	–

IDEAS CLAVES EN TERMINOLOGIA COMUN DE ENVEJECIMIENTO

Causas de la degradación

- **Condiciones de servicio** son todas las condiciones actuales que influyen sobre un SEC. Ellas comprenden **condiciones de operación** (incluyendo **normales** y **condiciones inducidas por error** asi como transitorios anticipados) y condiciones de accidentes.

- **Condiciones de diseño** son condiciones hipotéticas generalmente especificadas para incluir un margen conservador de condiciones esperadas actuales de servicio.

Vida

- **Vida de servicio** es el periodo actual de servicio útil de un SEC. Esta puede ser diferente de la vida esperada de servicio, por ejemplo la **vida de diseño**.

- La **edad** de un SEC (medida desde su tiempo de fabricación) puede diferir de su **tiempo en servicio** (medida desde la operacion inicial del SEC)

Degradación/Envejecimiento

- **Degradación** es gradual (envejecimiento) o inmediata (no-envejecimiento).

- **Degradación por envejecimiento** es producido por **condiciones de operación**, incluyendo **condiciones ambientales** tales como temperatura y radiación asi como **condiciones funcionales** tal como movimiento relativo entre las partes. Condiciones de operación producen **agentes de envejecimiento normales** o **agentes de envejecimiento inducidos por error**.

- **Sucesos bases de diseño** incluyen transitorios anticipados durante la operación de planta (los cuales pueden contribuir a **envejecimiento**) y accidentes bases de diseño y terremotos (los cuales producen **degradación** inmediata no envejecimiento.

- Una **condición degradada** es marginalmente aceptable pero no es un **fallo**. En un **fallo degradada**, **los criterios de aceptación** de condicion o comportamiento no se cumplen, pero existe un funcionamiento parcial. En un **fallo completo** no funciona.

Fallo

- **Fallo** es usualmente producido por una cadena sequencial de causas, no una sola causa. **Desgaste** es una fallo cuya ultima causa es un **mecanismo de envejecimiento**. La **causa raíz** podría no ser ese mecanismo de envejecimiento.

- **Envejecimiento prematuro** podrá ser causado por un **fallo** en servicio de un SEC. El término **envejecimiento acelerado** debe ser reservado a **envejecimiento artificial** usualmente desarrollado en un laboratorio.

- **Analisis de fallo** identifica **causas de fallo**, los **mecanismos de fallo** y el **modo del fallo**. Cada uno de estos terminos tiene un significado diferente. La **causa raíz** de una **degradacion**
  por envejecimiento inducida por error y su fallo no es envejecimiento sino un error humano.

Mantenimiento/Evaluación de la condición

- **Mantenimiento** es un término amplio que incluye **mantenimiento correctivo** y **mantenimiento preventivo**. **El mantenimiento** puede ser implementado por personal de mantenimiento, ingeniería u operaciones. **El mantenimiento preventivo** incluye **mantenimiento predictivo** tal como **inspección**, **ensayo**, y **control de condición**.

- **Reparación** es desarrollada solo en un SEC que ha falllado; **reconstrucción** es desarrollada solo en un SEC que no ha fallado. Una **revisión general** es una reparación o reconstrución extensiva.

Cuadro 1. **Relaciones entre los categorios de términos**

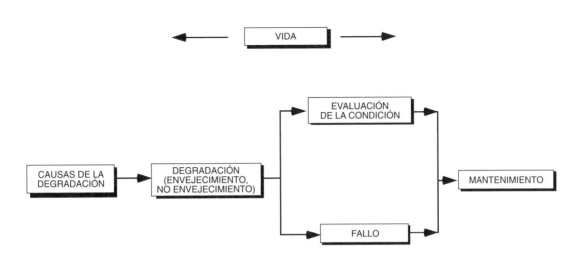

Nota : La evaluación de la condición es la parte predictiva del mantenimiento.

Cuadro 2. **Relaciones entre tipos de condiciones de servicio**

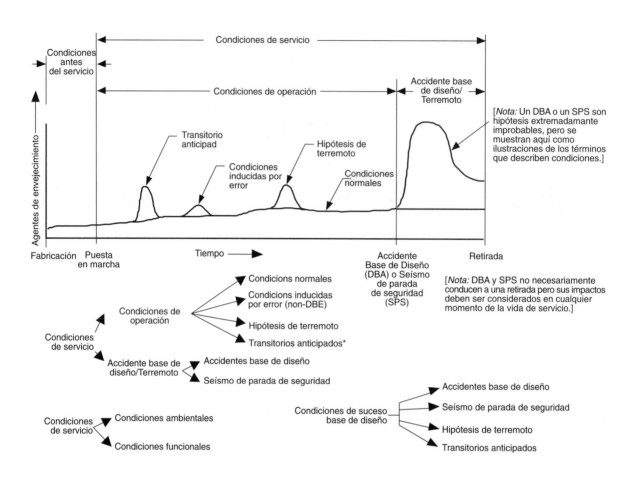

* Nótese que los transitorios anticipados, aunque estén previsitos en el diseño, pueden ser inducidos por error. Las condiciones de diseño comprenden todas las condiciones de servicio con excepción de las condiciones no DBE inducidas por error.

Cuadro 3. **Relaciones entre términos de envejecimiento**

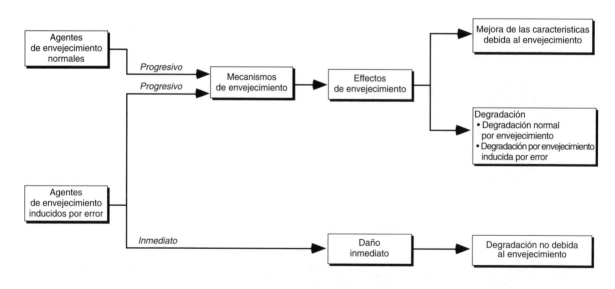

Cuadro 4. **Relaciones entre términos referidos a sucesos reales ordenados en secuencia linea**

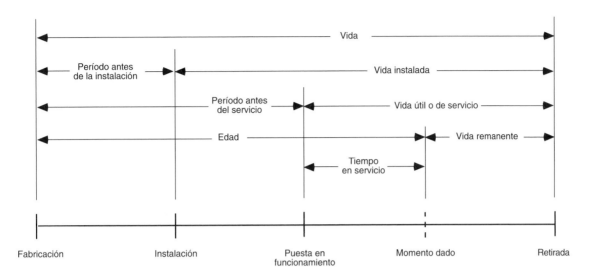

Nota: Estos términos se refieren a hitos del pasado o del futuro de un sistema, estructuras o componente.

Cuadro 5. **Relaciones entre términos de predicción de vida de diseño**

A. Predicción inicial de vida de diseño

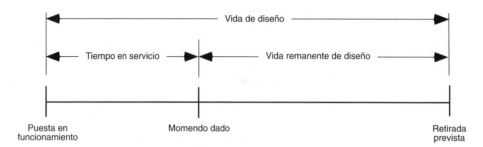

B. Prolongación posible de la vida de diseño
tras nuevo cálculo de vida de diseño (discrecional)

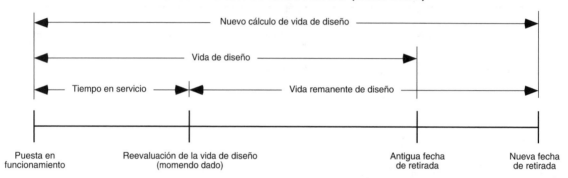

C. Disminución posible de la vida de diseño
tras nuevo cálculo de vida de diseño (discrecional)

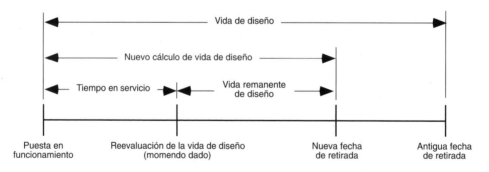

Cuadro 6. **Relación entre términos de fallos**

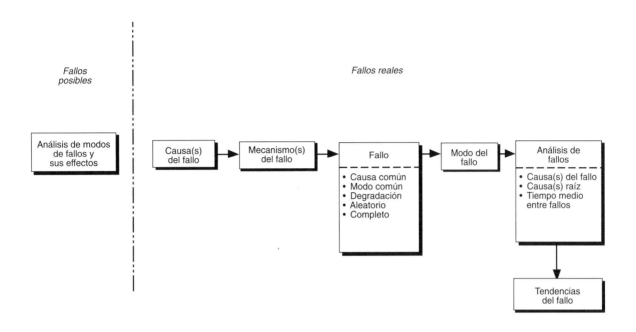

Cuadro 7. **Ejemplos de causas, mecanismos y modos**

Definiciones de causa raíz y desgaste

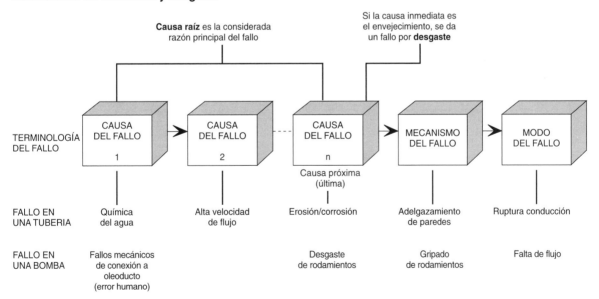

Cuadro 8.   **Terminología de mantenimiento**

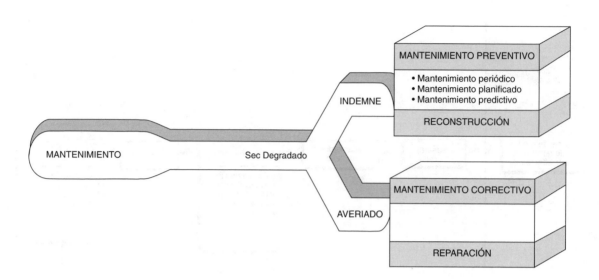

Cuadro 9. **Relaciones entre términos de mantenimiento**

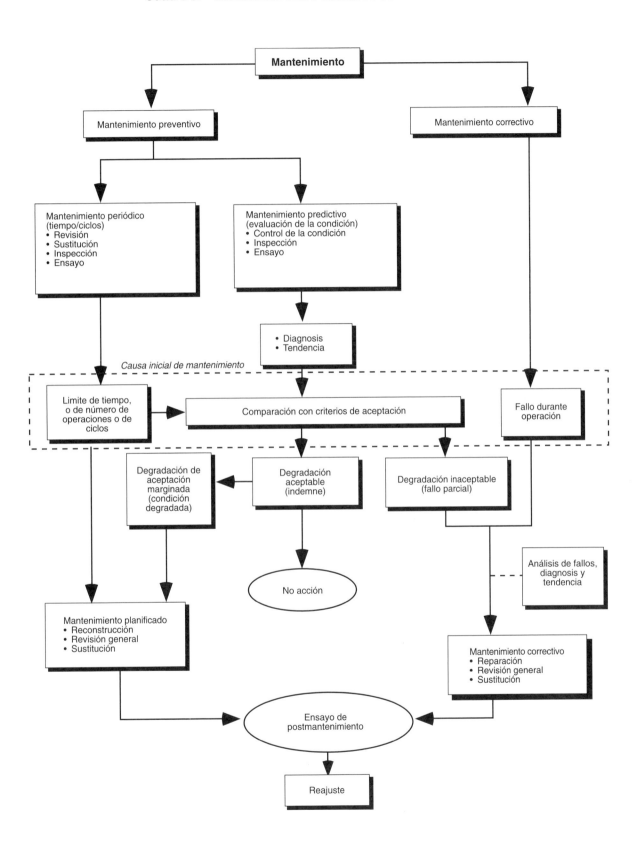

ГЛОССАРИЙ СТАРЕНИЯ АТОМНЫХ ЭЛЕКТРОСТАНЦИЙ

ГЛОССАРИЙ, ПОЛЕЗНЫЙ ДЛЯ ПОНИМАНИЯ И УПРАВЛЕНИЯ СТАРЕНИЕМ СИСТЕМ, СТРУКТУР И КОМПОНЕНТОВ АТОМНЫХ ЭЛЕКТРОСТАНЦИЙ

В ЧЕМ НЕОБХОДИМОСТЬ КАРМАННОГО ГЛОССАРИЯ?

С возрастанием срока службы АЭС, находящихся в эксплуатации, вопросу возможного непонимания терминов, относящихся к ухудшению состояния систем, структур и отдельных компонентов АЭС вследствие старения уделяется все больше внимания. Общая терминология старения была разработана для улучшения понимания явлений старения, обеспечения правильного документирования данных об отказах на АЭС, а также для развития единого толкования стандартов и норм безопасности, которые относятся к вопросам старения.

Эта терминология должна быть полезной в области управления процессами старения и сроком службы. Управление сроком службы может способствовать снижению стоимости эксплуатации, технического обслуживания и ремонта, а также может обеспечить возможность обоснования продления срока службы АЭС от 40 до 60 лет. Что еще более важно, эффективное управление вопросами старения вносит вклад в поддержание адекватных запасов безопасности оборудования АЭС.

Признавая важность четкого взаимодействия в этих областях, представители различных отраслей ядерной промышленности разработали единый словарь выражений, относящихся к процессам старения.

Имея в виду пользу, которая может быть получена от использования общей терминологии старения, ее использование рекомендуется в технических документах, отчетах об отказах, исследовательских отчетах, в новых регулирующих документах и другой документации, относящейся к вопросам старения АЭС. Документация должна разъяснять, что общая терминология старения применяется во всех случаях, за исключением отдельно указанных. Соответствующие исключения могут быть сделаны для тех случаев, в которых автор предпочитает использование определений из уже существующих норм и стандартов.

Основными выгодами от использования общей терминологии старения являются следующие:

- улучшение отчетности и толкования данных об ухудшении состояния оборудования АЭС (систем, структур и компонентов) и его отказах, включая точное определение их причин;

- улучшение толкования и согласованность терминов с кодами, стандартами и нормами, относящимися к процессам старения АЭС.

В чем необходимость глоссария?

Агентство по ядерной энергии (АЯЭ) совместно с Комиссией Европейских Сообществ (КЭС) и Международным Агенством по Атомной Энергии (МАГАТЭ) опубликовало этот глоссарий как удобный справочник, чтобы способствовать широкому использованию общей терминологии старения и облегчить ее применение. Основной целью является обеспечение персонала АЭС (а также других лиц, имеющих дело со старением оборудования) общим набором терминов, которые имеют единое во всех отраслях промышленности значение, чтобы облегчить дискуссии между экспертами из разных стран.

В каждом языковом разделе термины перечисляются в алфавитном порядке и имеют последовательную нумерацию. Эта нумерация повторяется в секции английского языка, позволяя таким образом осуществить перекрестные ссылки между всеми используемыми языками. Тире ("-") в колонках с номерами для ссылок означает, что в данном языке соответствующий термин отсутсвует. Следует обратить внимание на то, что тире используется только в колонках с синонимами.

На каждом языке глоссарий начинается с обзора всех терминов, сгруппированных в шесть категорий. За обзором следуют алфавитный перечень терминов, определений и некоторые примеры. Последние страницы каждого раздела содержат диаграммы и перечень ключевых идей для прояснения терминологии.

Выражение признательности

Данный глоссарий во многом близок публикации Исследовательского института электроэнергетики США (EPRI) (BR-101747), которому мы признательны за поддержку и практические рекомендации при подготовке этой международной версии глоссария. Глоссарий EPRI был подготовлен при содействии ряда электрических комнаний США, Института ядерной энергетики, Комиссии по ядерному регулированию США и национальных лабораторий США.

Основные общие термины старения по категориям

ухудшение свойств	
причины ухудшения свойств	ухудшение свойств / старение
условие (условия) • рабочие условия • предэксплуатационные условия • условия окружающей среды (окружающие условия) • функциональные условия • эксплуатационные условия • нормальные условия • условия, вызванные ошибкой • исходное событие проектной аварии • условия исходных событий для проектной аварии • проектные условия фактор воздействия • фактор воздействия при нормальных условиях • фактор воздействия, вызванный ошибкой • фактор воздействия от исходного события проектной аварии	характеристика состояние • ухудшившееся состояние старение • естественное старение • преждевременное старение • нормальное старение • искусственное старение • ускоренное старение • моделирование старения механизм старения эффекты старения • совокупные эффекты • одновременные эффекты • синергические эффекты ухудшение свойств • ухудшение свойств вследствие старения • ухудшение свойств при нормальном старении • ухудшение свойств при старении, вызванное ошибкой оценка старения

жизненный цикл	
жизненный цикл	отказ
возраст период эксплуатации жизненный цикл • срок службы • ресурс • остаточный ресурс • проектный ресурс • остаточный проектный ресурс • установленный ресурс демонтаж	отказ • отказ с частичной потерей функции • полный отказ • случайный отказ • отказы по общей причине • зависимый отказ общего вида • износ (отказ вследствие старения) причина отказа • коренная причина • механизм отказа • вид отказа анализ отказов • оценка отказа • анализ видов и последствий отказов • прогнозирование отказа среднее время между отказами

управление старением	
техническое обслуживание и ремонт	оценка условий
управление старением управление жизненным циклом техническое обслуживание и ремонт • планово-предупредительное техническое обслуживание и ремонт • периодическое техническое обслуживание и ремонт • плановое техническое обслуживание и ремонт • исправительное техническое обслуживание и ремонт ремонтно-восстановительные работы (ремонт) восстановление (реконструкция) капитальный ремонт замена техническое обслуживание испытание после технического обслуживания и ремонта доработка (переделка)	профилактическое техническое обслуживание и ремонт (или ТО и ремонт по прогнозу) эксплуатационная инспекция эксплуатационное испытание эксплуатационный надзор / контроль надзорные требования контроль состояния показатель состояния функциональный показатель испытание диагностика критерий приемлемости

		англ	фран	нем	испан
1	**анализ видов и последствий отказов** — систематический процесс определения и документирования возможных видов отказов и их влияния на СКК[1]	48	4	42	7
2	**анализ коренных причин** — синоним *анализа отказов*	97	2	20	5
3	**анализ отказов** — систематический процесс определения и документирования вида, механизма, причин и коренной причины отказа СКК	43	3	20	6
4	**вид отказа** — характер развития отказа или состояние, при котором наступает отказ СКК	47	87	87	75

• *Примеры*: непосадка или заклинивание в открытом состоянии (клапана); короткое замыкание на землю (кабеля); заклинивание подшипников (двигателя); утечка (через клапан, корпус или защитную оболочку), остановка или перекрытие потока (в трубопроводе или клапаном), непоступление сигнала, который опускает управляющие стержни (системы защиты реактора); трещина или поломка (конструкции)

		англ	фран	нем	испан
5	**возраст** — промежуток времени от момента изготовления СКК до данного момента	3	1	4	38
6	**восстановление (реконструкция)** — плановые меры, принимаемые с целью улучшения состояния не отказавших СКК	86	94	59	78
7	**выход из строя** — синоним *полного отказа*	13	–	79	8
8	**демонтаж** — окончательный вывод СКК из эксплуатации	94	90	28	81
9	**диагностика** — обследование и оценка данных для определения состояния СКК или причин, вызвавших их настоящее состояние	36	35	37	37
10	**диагностическая оценка** — синоним *диагностики*	37	62	37	–
11	**доработка (переделка)** — исправление ненадлежащим образом выполненного изготовления, монтажа или технического обслуживания	95	91	58	77

---

1. СКК = система, конструкция или компонент

		англ	фран	нем	испан
12	**естественное старение** – старение СКК, которое происходит в предшествующих рабочим условиях и в самих рабочих условиях, включая условия, вызванные ошибками	64	103	60	49
13	**жизненный цикл** – время с момента изготовления до демонтажа СКК	56	36	53	90
14	**зависимый отказ общего вида** – два или большее число случаев отказа одинакового вида или характера вследствие одной причины	17	26	19, 77	58
15	**замена** – удаление СКК или их частей, свойства которых не подверглись ухудшению, ухудшились или которые оказались отказавшими (потерявшими работоспособность) и монтаж на их место других СКК, которые могут функционировать, обеспечивая соблюдение первоначальных критериев приемлемости	92	93	29	85
16	**износ (отказ вследствие старения)** – отказ вследствие действия механизма старения	111	99	17	35
17	**искусственное старение** – моделирование эффектов естественного старения у СКК посредством приложения факторов воздействия, имитирующих предэксплуатационные и рабочие условия станции, которые, однако, могут отличаться от них по интенсивности, продолжительности и способу приложения	12	102	51	48
18	**исправительное техническое обслуживание и ремонт** – меры, которые восстанавливают посредством ремонтных работ, капитального ремонта или замены способность отказавших (потерявших работоспособность) СКК функционировать, обеспечивая соблюдение критериев приемлемости	24	80	49	68
19	**испытание** – наблюдение или измерение показателей состояния в контролируемых условиях с целью проверки того, что в данный момент СКК удовлетворяют критериям приемлемости	108	54	68, 78	43

		англ	фран	нем	испан
20	**испытание после технического обслуживания и ремонта** – испытание после технического обслуживания и ремонта, проводимое с целью убедиться в том, что техническое обслуживание и ремонт были проведены правильно и что СКК могут функционировать, обеспечивая соблюдение критериев приемлемости	77	55	69	44
21	**исходное событие проектной аварии** – событие, которое определяется для обоснования надежной работы и – посредством детерминированного анализа безопасности – связанных с безопасностью функций СКК; события включают предусмотренные проектом переходные режимы, проектные аварии, внешние события и природные явления	29	66	26	84
22	**капитальный ремонт** – крупный ремонт, восстановление (реконструкция) или оба этих вида работ	73	96	81	83
23	**контроль** – постоянный контроль состояния станции во время эксплуатации или остановов	104	–	83	99
24	**контроль состояния** – наблюдение, измерение или прогнозирование[2] показателей состояния или функциональных показателей относительно некоторого независимого параметра (обычно времени или циклов) для получения вывода о способности СКК функционировать в настоящее время или в будущем, обеспечивая соблюдение критериев приемлемости	22	97	101	28
25	**коренная причина** – основная(е) причина(ы) наблюдаемого состояния СКК, которая(ые) после ее(их) устранения исключает(ют) повторное возникновения этого состояния	96	7	39	11
26	**критерий приемлемости** – определенное предельное значение функционального показателя или показателя состояния, используемое для оценки способности СКК выполнять проектную функцию	2	22	3	29

---

2. Определение тренда в наблюдениях или измерениях

		англ	фран	нем	испан
27	**механизм отказа** — физический процесс, который приводит к отказу · Примеры: растрескивание охрупченной изоляции кабеля (обусловлено старением); появление предмета, препятствующего потоку (не обусловлено старением)	46	85	22	74
28	**механизм старения** — конкретный процесс, который постепенно изменяет характеристики СКК со временем или в результате использования · Примеры: отвердение, износ, усталость, ползучесть, эрозия, микробиологическое обрастание, коррозия, охрупчивание, химические или биологические реакции	10	86	11	73
29	**моделирование старения** — моделирование эффектов естественного старения у СКК посредством применения любого сочетания искусственного и естественного старения	4	60	89	52
30	**надзорные испытания** — синоним *эксплуатационного надзора, надзорных требований и эксплуатационных испытаний* (термин используется только в контексте конкретных вопросов регулирования или юридических аспектов)	106	56	–	–
31	**надзорные требования** — испытание, калибровка или инспекция, проводимые с тем, чтобы обеспечить поддержание необходимого качества систем и компонентов, эксплуатацию с соблюдением пределов безопасности и соблюдение ограничительных условий эксплуатации (термин используется только в контексте конкретных вопросов регулирования или юридических аспектов)	105	89	84	80
32	**неисправность** — синоним отказа *(потери работоспособности)*	62	28	15, 86	–
33	**нормальное старение** — естественное старение в результате воздействия условий, предшествующих рабочим, или самих рабочих условий при отсутствии воздействий, вызываемых ошибками	65	104	62	50

		англ	фран	нем	испан
34	**нормальные условия** – эксплуатационные условия надлежащим образом спроектированных, изготовленных, смонтированных, эксплуатируемых и проходящих техническое обслуживание СКК, за исключением условий исходных событий проектной аварии	67	19	61	26
35	**нормальные эксплуатационные условия** – синоним *нормальных условий*	68	20	61	20
36	**одновременные эффекты** – суммарные эффекты от факторов воздействия, действующих одновременно	101	50	74	42
37	**остаточный проектный ресурс** – время от данного момента до запланированного демонтажа СКК	87	42	71	96
38	**остаточный ресурс** – реальный интервал времени от данного момента до демонтажа СКК	88	44	72	95
39	**отказ** – неспособность или потеря способности СКК функционировать, обеспечивая соблюдение критериев приемлемости	42	23	15, 86	54
40	**отказ с частичной потерей функции** – отказ, при котором функциональный показатель не отвечает критерию приемлемости, однако проектная функция утеряна не полностью	27	71	18	59
41	**отказы по общей причине** – два или большее число случаев отказа вследствие одной причины	16	27	16, 77	57
42	**оценка жизненного цикла** – синоним *оценки старения*	57	64	54	–
43	**оценка отказа** – вывод, полученный в результате анализа отказа (потери работоспособности)	45	63	21	53
44	**оценка старения** – оценка соответствующей информации в целях определения эффектов старения, оказывающих влияние на способность СКК функционировать в настоящее время или в будущем, обеспечивая соблюдение критериев приемлемости	6	65	9	89

		англ	фран	нем	испан
45	**периодическое техническое обслуживание и ремонт** − форма планово-предупредительного технического обслуживания и ремонта, заключающаяся в обслуживании, замене деталей, контроле или проверке через заранее установленные интервалы в шкале календарного времени, периода эксплуатации или количества циклов	75	81	66	69
46	**период эксплуатации** − время от начала эксплуатации СКК до данного момента	109	47	36	87
47	**планово-предупредительное техническое обслуживание и ремонт** − меры, посредством которых обнаруживается, предотвращается или ограничивается ухудшение свойств действующих СКК в целях сохранения или продления их полезного срока эксплуатации (службы) путем контроля и ограничения процесса ухудшения свойств и отказов (потери работоспособности) до приемлемого уровня; существует три типа планово-предупредительного технического обслуживания: периодическое, профилактическое (по прогнозу) и плановое	82	82	91	72
48	**плановое техническое обслуживание и ремонт** − форма планово-предупредительного технического обслуживания и ремонта, заключающаяся в восстановлении (реконструкции) или замене, которые запланированы и выполняются до наступления прогнозируемого неприемлемого ухудшения характеристик СКК	76	84	67	70
49	**показатель работоспособности** − синоним *функционального показателя*	74	76	44	63
50	**показатель состояния** − характеристика, которую можно контролировать, измерять или прогнозировать для косвенного или прямого определения способности СКК функционировать в настоящее время или в будущем, обеспечивая соблюдение критериев приемлемости	21	75	100	62
51	**полный отказ** − отказ, при котором происходит полная потеря функции	18	25	79	56

		англ	фран	нем	испан
52	**предэксплуатационные условия** — реальное физическое состояние СКК или воздействие на них до начала эксплуатации (например, изготовление, хранение, транспортирование, монтаж, предэксплуатационные испытания)	81	10	33, 92	16
53	**преждевременное старение** — старение в процессе эксплуатации, которое происходит с более высокой скоростью, чем ожидалось	80	105	93	51
54	**причина отказа** — обстоятельства в процессе проектирования, изготовления, испытания или использования, которые привели к отказу (потере работоспособности)	44	6	23	10
55	**прогнозирование отказа** — регистрация, анализ и экстраполирование отказов (потери работоспособности) СКК в процессе эксплуатации относительно некоторого независимого параметра (обычно времени или циклов)	49	58	80	86
56	**прогнозирование состояния** — синоним *контроля состояния*	23	57	101	—
57	**проектные условия** — рабочие условия, определенные для СКК в соответствии с существующими правилами и нормами, которые обычно включаются в технические спецификации (обычно применяют консервативный подход: оценка с запасом ожидаемых эксплуатационных параметров)	32	18	25	18
58	**проектный ресурс** — период, в течение которого ожидается, что СКК будет функционировать, обеспечивая соблюдение критериев приемлемости	33	41	27	91
59	**профилактическое техническое обслуживание (ТО) и ремонт (или ТО и ремонт по прогнозу)** — форма планово-предупредительного технического обслуживания и ремонта, проводимого на непрерывной основе или через интервалы, устанавливаемые с учетом наблюдаемого состояния с целью контроля, диагностики или прогнозирования функциональных показателей или показателей состояния СКК; результаты позволяют определить функциональную способность в данный момент времени или в будущем, или установить характер и график проведения планового технического обслуживания и ремонта	79	83	90	71

		англ	фран	нем	испан
60	**рабочие условия** − реальное физическое состояние или воздействие в течение срока службы (эксплуатации) СКК, включая эксплуатационные условия (нормальные или вызванные ошибкой), условия, возникающие в случае исходного события проектной аварии, и условия, возникающие после исходного события проектной аварии	98	14	40, 52	21
61	**ремонтно-восстановительные работы (ремонт)** − меры, направленные на возвращение отказавших (потерявших работоспособность) СКК в приемлемое состояние	91	95	70	79
62	**ресурс** − реальное время от начала эксплуатации до демонтажа СКК	99	38	41, 65, 75	92
63	**синергические эффекты** − часть изменений характеристик СКК, обусловленная исключительно взаимодействием одновременно действующих факторов воздействия, в отличие от суммарных изменений, вызванных наложением всех факторов воздействия, действующих независимо	107	51	76	41
64	**случайный отказ** − любой отказ, причины или механизмы которого, или то и другое не позволяют предсказать время их возникновения	84	24	98	55
65	**совокупные эффекты** − суммарные изменения характеристик СКК, вызванные двумя или большим числом факторов воздействия	15	48	50	39
66	**состояние** − состояние или уровень характеристик СКК, которые могут оказывать влияние на их способность выполнять проектную функцию	20	−	99	13
67	**среднее время между отказами** − среднее арифметическое времени эксплуатации в периоды между отказами элемента	63	88	57	88
68	**срок службы** − время с момента монтажа до демонтажа СКК	55	37	47	94
69	**старение** − общий процесс, при котором характеристики СКК постепенно изменяются со временем или в результате использования	5	106	5	46

		англ	фран	нем	испан
70	**техническое обслуживание** – плановые меры (включая чистку, регулировку, калибровку и замену расходных материалов), которые поддерживают или продлевают полезный срок службы (эксплуатации) СКК	100	52	94	82
71	**техническое обслуживание и ремонт** – сочетание прямых и вспомогательных мер, посредством которых обнаруживается, предотвращается или ограничивается ухудшение свойств действующих СКК или восстанавливаются до приемлемого уровня проектные функции отказавших (потерявших работоспособность) СКК	61	79	48, 95	67
72	**управление жизненным циклом** – совокупность мер по управлению старением и экономическому планированию в целях: 1) оптимизации эксплуатации, технического обслуживания и срока службы СКК; 2) поддержания приемлемого уровня работоспособности и безопасности; 3) получения максимальной прибыли на инвестированный капитал в течение срока службы станции	59	72	55	60
73	**управление старением** – технические меры, эксплуатационные меры и техническое обслуживание, осуществляемые в целях поддержания в приемлемых пределах ухудшения свойств СКК при их старении и износе • Примеры технических мер: проектирование, аттестация и анализ отказов (потери работоспособности) • Примеры эксплуатационных мер: контроль (надзор), выполнение эксплуатационных мероприятий с соблюдением определенных пределов, измерение параметров окружающей среды	9	74	10	61
74	**ускоренное старение** – искусственное старение, при котором моделирование естественного старения за короткое время приближенно воспроизводит эффекты старения, соответствующих более длительным рабочим условиям (см. также *преждевременное старение*)	1	101	34	47
75	**условие (условия)** – окружающее физическое состояние или действие, которое может оказывать влияние на СКК	19	8	31	12

		англ	фран	нем	испан
76	**условия, вызванные ошибкой** — неблагоприятные предэксплуатационные или рабочие условия, вызванные ошибками при проектировании, изготовлении, монтаже, эксплуатации или техническом обслуживании	40	17	14, 46	25
77	**условия исходных событий проектной аварии** — эксплуатационные условия, вызываемые исходными событиями проектной аварии	30	11	32	23
78	**условия окружающей среды (окружающие условия)** — окружающие (или внешние) физические условия, в которых находятся СКК	38	9	85	15
	• Примеры: температура, радиация и влажность в защитной оболочке (контейнменте) во время нормальной эксплуатации или аварий				
79	**установленный ресурс** — время, в течение которого, как было доказано на основе испытания, анализа или эксплуатационного опыта, СКК будут способны функционировать, обеспечивая соблюдение критериев приемлемости, в определенных эксплуатационных условиях при сохранении способности выполнения своих функций безопасности при проектной аварии или проектном землетрясении	83	40	45	76
80	**ухудшение свойств** — внезапное или постепенное ухудшение характеристик СКК, которое может отрицательно сказаться на их способности функционировать, обеспечивая соблюдение критериев приемлемости	25	29	2, 88	30
81	**ухудшение свойств вследствие старения** — эффекты старения, которые могут ухудшить способность СКК функционировать, обеспечивая соблюдение критериев приемлемости	7	31	7, 8	31
	• Примеры: уменьшение диаметра вращающегося вала в результате износа, потеря прочности материала в результате усталости или термического старения, распухание герметиков, потеря электрической прочности диэлектрика или растрескивание изоляции				

		англ	фран	нем	испан
82	**ухудшение свойств при нормальном старении** – ухудшение свойств при старении в нормальных условиях	66	33	66	33
83	**ухудшение свойств при старении, вызванное ошибкой** – ухудшение свойств при старении, к которому привели условия, вызванные ошибкой	39	32	13	32
84	**ухудшившееся состояние** – ограниченно приемлемое состояние не потерявшего работоспособность СКК, которое может привести к решению провести плановое техническое обслуживание и ремонт	26	61	1	14
85	**фактор воздействия** – фактор или активное воздействие, которые возникают в результате влияния предэксплуатационных и рабочих условий и которые могут вызвать внезапное ухудшение свойств СКК или ухудшение свойств вследствие старения · Примеры: нагрев, излучение, влажность, пар, химические вещества, давление, вибрация, сейсмическое движение, электрические колебания, механическое циклирование	102	67	30	1
86	**фактор воздействия, вызванный ошибкой** – фактор воздействия, который возникает в результате условий, вызванных ошибкой, и может приводить к внезапному ухудшению свойств или ухудшению свойств вследствие старения в большей степени, чем нормальные факторы воздействия	41	69	12	2
87	**фактор воздействия от исходного события проектной аварии** – фактор воздействия, который возникает в результате исходного события проектной аварии и может вызвать внезапное ухудшение свойств или ухудшение свойств вследствие старения в большей степени, чем нормальные факторы воздействия	31	68	24	4
88	**фактор воздействия при нормальных условиях (нормальный фактор воздействия)** – фактор воздействия, который возникает при нормальных условиях и может вызвать действие механизмов старения и эффекты старения у СКК	69	70	64	3
89	**функциональный показатель** – показатель состояния, который непосредственно отражает способность СКК функционировать в данный момент, обеспечивая соблюдение критериев приемлемости	51	77	44	64

		англ	фран	нем	испан
90	**функциональные условия** – условия, оказывающие влияние на СКК в результате выполнения проектных функций (работы системы или компонента, нагрузки на конструкцию)	50	16	43	24
	• Примеры: для запорного клапана – цикличность срабатывания и стук; для предохранительного клапана корпуса реактора – давление теплоносителя реактора, высокие скорости потока и повышение температуры от теплоносителя реактора				
91	**характеристика** – свойство, качество, показатель или признак СКК (например: форма; размер; вес; показатель состояния; функциональный показатель; работоспособность; механическое, химическое или электрическое свойство)	14	5	38	9
92	**эксплуатационная инспекция** – методы и действия, направленные на обеспечение структурной целостности и способности выдерживать давление связанных с безопасностью компонентов АЭС	52	78	73,97	66
93	**эксплуатационное испытание** – испытание, проводимое в целях определения эксплуатационной готовности компонента или системы	54	53	96	45
94	**эксплуатационный надзор** – наблюдение или измерение показателей состояния или функциональных показателей с целью проверки того, что СКК в данный момент могут функционировать, обеспечивая соблюдение критериев приемлемости	103	98	82	65
95	**эксплуатационные условия** – рабочие условия, включая нормальные условия и условия, вызванные ошибкой, существующие до возникновения проектной аварии или проектного землетрясения	70	12	35	19
96	**эффекты старения** – совокупные изменения характеристик СКК, которые возникают со временем или в результате использования и являются следствием действия механизмов старения	8	49	6	40
	• Примеры: отрицательные эффекты – см. *ухудшение свойств при старении;* положительные эффекты – повышение прочности бетона в результате твердения; снижение вибрации вследствие приработки вращающихся деталей				

КЛЮЧЕВЫЕ ПОНЯТИЯ В ОБЩЕЙ ТЕРМИНОЛОГИИ ПО СТАРЕНИЮ

ПРИЧИНЫ УХУДШЕНИЯ СВОЙСТВ

- **Рабочие условия** – это все реальные условия, которые оказывают влияние на СКК. В их число входят **эксплуатационные условия** (включая **нормальные условия и условия, вызванные ошибкой,** а также предусмотренные проектом переходные процессы) и аварийные условия.

- **Проектные условия** – это гипотетические условия, обычно определяемые с запасом по отношению к ожидаемым реальным рабочим условиям.

ЖИЗНЕННЫЙ ЦИКЛ

- **Ресурс** – это реальное время, в течение которого СКК обеспечивают полезную эксплуатацию. Он может отличаться от ожидаемого ресурса, т.е. **проектного ресурса.**

- **Возраст** СКК (измеряемый от времени их изготовления) может отличаться от его **времени эксплуатации** (измеряемого от начала эксплуатации СКК).

УХУДШЕНИЕ СВОЙСТВ/СТАРЕНИЕ

- **Ухудшение свойств** происходит постепенно (старение) или внезапно (не связано со старением).

- **Ухудшение свойств при старении** вызывается **эксплуатационными условиями,** включая такие **условия окружающей среды (окружающие условия),** как температура и излучение, а также такие **функциональные условия,** как относительное движение составляющих частей. Эксплуатационные условия являются причиной возникновения **факторов воздействия при нормальных условиях** или **факторов воздействия, вызванных ошибкой.**

- **Исходные события проектной аварии** включают предусмотренные проектом переходные режимы во время эксплуатации станции (которые могут способствовать **старению)** и проектные аварии и землетрясения (которые вызывают внезапное, не связанное со старением **ухудшение свойств).**

- **Ухудшившееся состояние** в некоторых пределах приемлемо, но не является

120

отказом. При отказе, обусловленном ухудшением свойств, не обеспечивается соблюдение **критериев приемлемости** в отношении состояния или работы, однако сохраняется частичное функционирование. При **полном отказе** функционирования не происходит.

ОТКАЗ

- **Отказ** обычно является следствием последовательной цепочки целого ряда причин, а не единственной причины. **Износ** – это отказ (потеря работоспособности), последней причиной которой является **механизм старения. Коренной причиной** может быть не механизм старения.

- **Преждевременное старение** может привести к **отказу** СКК во время эксплуатации. Термин **ускоренное старение** следует употреблять применительно к **искусственному старению,** обычно вызываемому в лабораторных условиях.

- **Анализ отказов** выявляет **причины отказов, механизм отказов и вид отказа.** Каждый из этих терминов имеет различное значение. **Коренной причиной ухудшения свойств при старении, вызванного ошибкой,** и отказов является не **старение,** а скорее ошибка человека.

ТЕХНИЧЕСКОЕ ОБСЛУЖИВАНИЕ И РЕМОНТ / ОЦЕНКА СОСТОЯНИЯ

- **Техническое обслуживание и ремонт** – это термин, имеющий широкое значение, который включает **исправительное техническое обслуживание и ремонт и планово-предупредительное техническое обслуживание и ремонт.** **Техническое обслуживание и ремонт** могут проводиться обслуживающим, инженерным или эксплуатационным персоналом. **Планово-предупредительное техническое обслуживание и ремонт** включает в себя **профилактическое техническое обслуживание и ремонт (или ТО и ремонт по прогнозу),** такое, как обследование, испытания и **контроль состояния**.

- **Ремонт** выполняется только для отказавших СКК; **восстановление (реконструкция)** осуществляется только для не отказавших СКК. **Капитальный ремонт** – это крупный ремонт и/или **восстановление (реконструкция).**

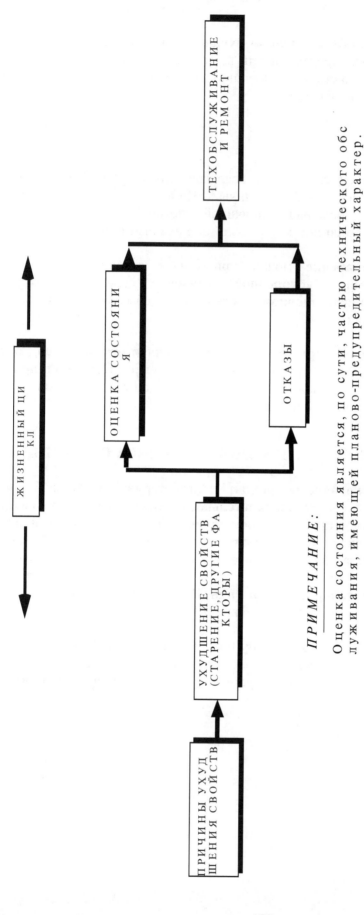

ЖИЗНЕННЫЙ ЦИКЛ

ПРИЧИНЫ УХУДШЕНИЯ СВОЙСТВ

УХУДШЕНИЕ СВОЙСТВ (СТАРЕНИЕ, ДРУГИЕ ФАКТОРЫ)

ОЦЕНКА СОСТОЯНИЯ

ОТКАЗЫ

ТЕХОБСЛУЖИВАНИЕ И РЕМОНТ

ПРИМЕЧАНИЕ:

Оценка состояния является, по сути, частью технического обслуживания, имеющей планово-предупредительный характер.

Диаграмма 1. Взаимоотношение категорий применяемых терминов

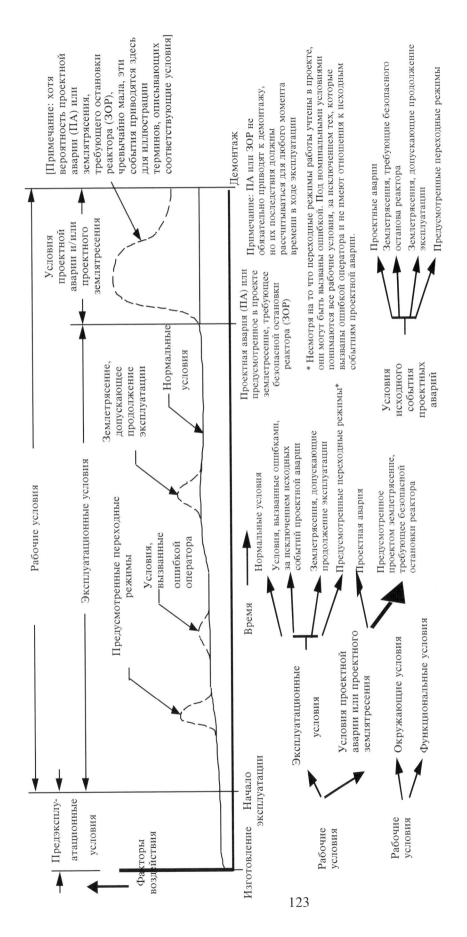

Диаграмма 2. Взаимоотношение видов рабочих условий.

123

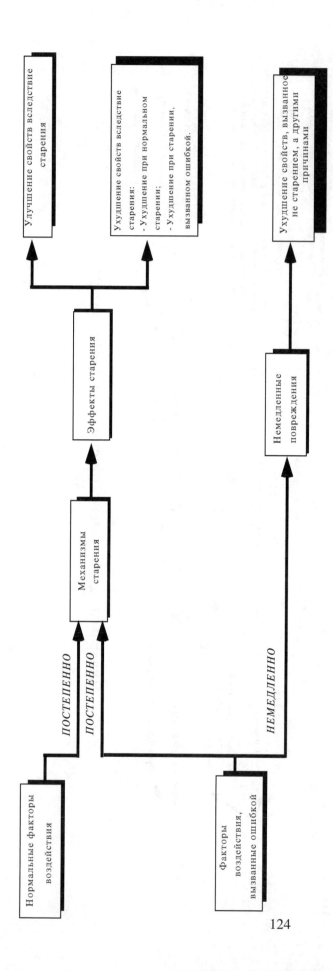

Диаграмма 3. Взаимоотношение терминов, относящихся к старению оборудования.

124

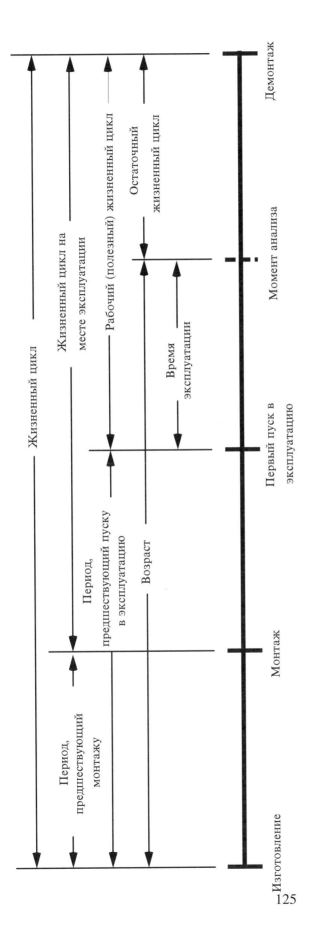

ПРИМЕЧАНИЕ: Данные термины служат для описания реальных этапов развития той или иной системы, структуры или компонента в прошлом или будущем. В скобках указаны синонимы.

Диаграмма 4. Взаимоотношение терминов, описывающих временную последовательность этапов реального жизненного цикла.

A. Первоначальное заключение о продолжительности номинального жизненного цикла

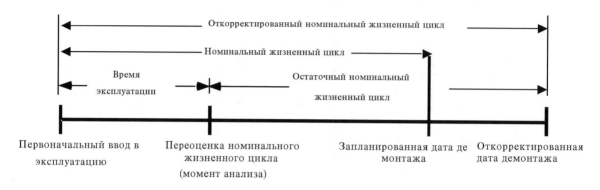

B. Возможное увеличение номинального жизненного цикла в результате корректировки (консервативной) предусмотренного номинального жизненного цикла

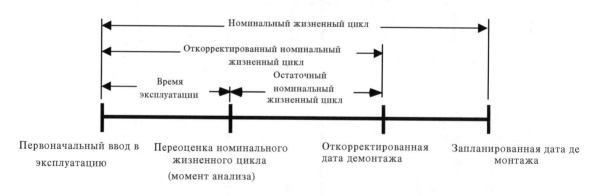

C. Возможное сокращение номинального жизненного цикла в результате корректировки (консервативной) предусмотренного номинального жизненного цикла

Диаграмма 5. Взаимоотношение терминов, относящихся к прогнозируемому номинальному жизненному циклу.

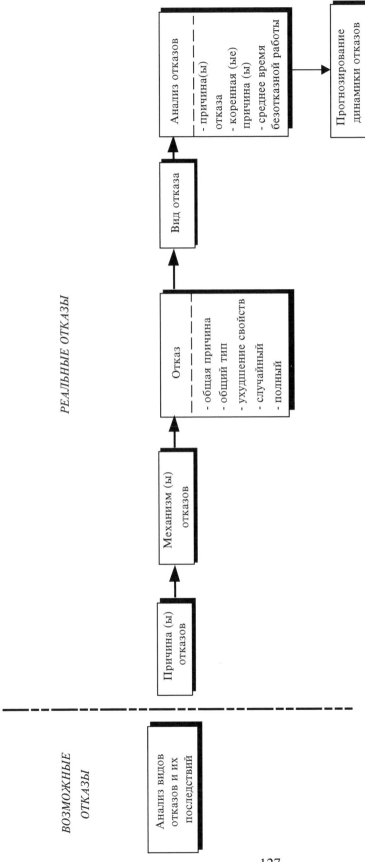

Диаграмма 6. Взаимоотношение терминов, описывающих отказы.

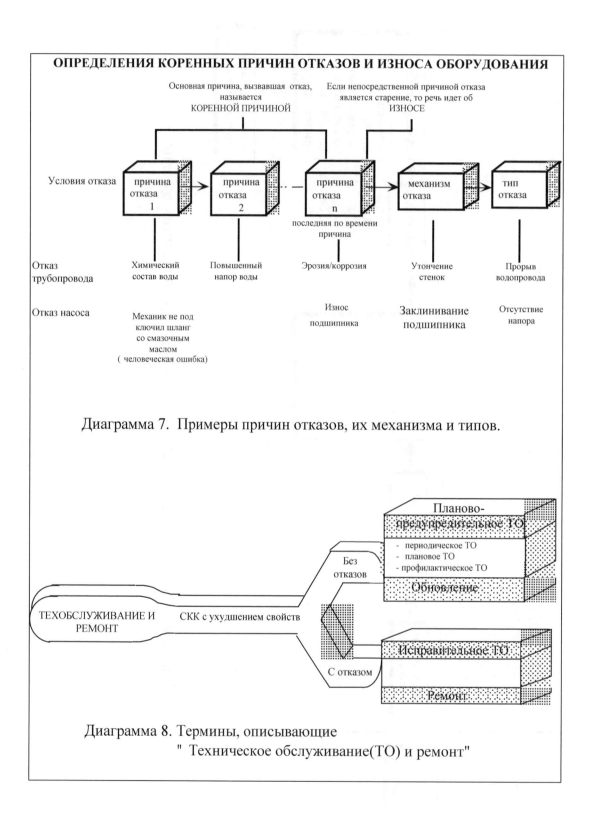

ОПРЕДЕЛЕНИЯ КОРЕННЫХ ПРИЧИН ОТКАЗОВ И ИЗНОСА ОБОРУДОВАНИЯ

Основная причина, вызвавшая отказ, называется
КОРЕННОЙ ПРИЧИНОЙ

Если непосредственной причиной отказа является старение, то речь идет об ИЗНОСЕ

Условия отказа — причина отказа 1 → причина отказа 2 → причина отказа n → механизм отказа → тип отказа

последняя по времени причина

Отказ трубопровода
Химический состав воды | Повышенный напор воды | Эрозия/коррозия | Утончение стенок | Прорыв водопровода

Отказ насоса
Механик не под ключил шланг со смазочным маслом (человеческая ошибка) | Износ подшипника | Заклинивание подшипника | Отсутствие напора

Диаграмма 7. Примеры причин отказов, их механизма и типов.

ТЕХОБСЛУЖИВАНИЕ И РЕМОНТ

СКК с ухудшением свойств

Без отказов

Планово-предупредительное ТО
- периодическое ТО
- плановое ТО
- профилактическое ТО
Обновление

С отказом

Исправительное ТО

Ремонт

Диаграмма 8. Термины, описывающие
" Техническое обслуживание(ТО) и ремонт"

128

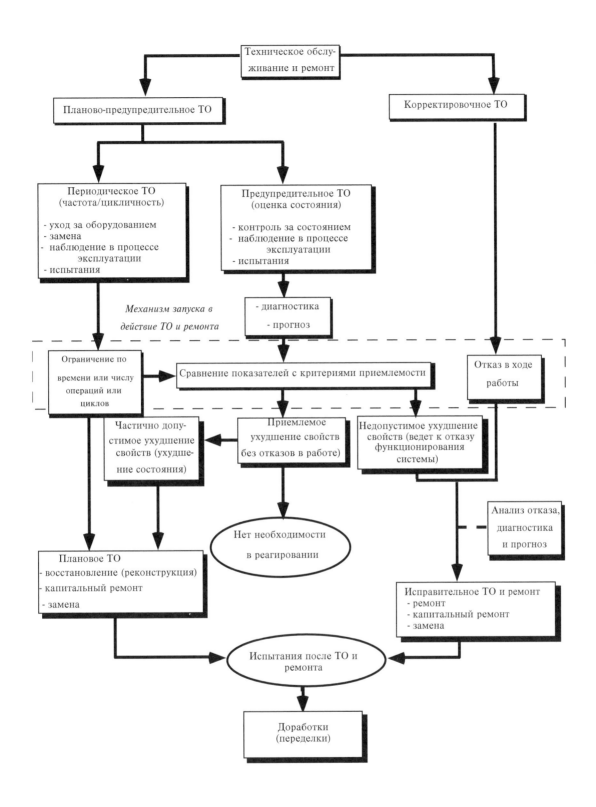

Диаграмма 9. Взаимоотношение терминов, связанных с техническим обслуживанием и ремонтом.

OECD PUBLICATIONS, 2, rue André-Pascal, 75775 PARIS CEDEX 16
PRINTED IN FRANCE
(66 1999 05 3 P) ISBN 92-64-05842-7 – No. 50557 1999